SYSTEMS ENGINEERING METHODS, DEVELOPMENTS AND TECHNOLOGY



# SYSTEMS ENGINEERING

## CONCEPTS, TOOLS AND APPLICATIONS

# SYSTEMS ENGINEERING METHODS, DEVELOPMENTS AND TECHNOLOGY

Additional books in this series can be found on Nova's website under the Series tab.

Additional e-books in this series can be found on Nova's website under the eBooks tab.

SYSTEMS ENGINEERING METHODS, DEVELOPMENTS AND TECHNOLOGY

# SYSTEMS ENGINEERING

## CONCEPTS, TOOLS AND APPLICATIONS

MARLENE HOPKINS
EDITOR

*New York*

## Library of Congress Cataloging-in-Publication Data

Names: Hopkins, Marlene, editor.
Title: Systems engineering : concepts, tools and applications / editors, Marlene Hopkins.
Other titles: Systems engineering (Hopkins)
Description: Hauppauge, New York : Nova Science Publishers, Inc., [2016] |
Series: Systems engineering methods, developments and technology | Includes bibliographical references and index.
Identifiers: LCCN 2016029798 (print) | LCCN 2016034266 (ebook) | ISBN 9781634857529 (softcover) | ISBN 9781634857710 (Ebook) | ISBN 9781634857710 ()
Subjects: LCSH: System analysis. | System design. | Systems engineering.
Classification: LCC TA168 .S869 2016 (print) | LCC TA168 (ebook) | DDC 620/.0042--dc23
LC record available at https://lccn.loc.gov/201602979

*Published by Nova Science Publishers, Inc. † New York*

# CONTENTS


| | | |
|---|---|---|
| **Preface** | | **vii** |
| **Chapter 1** | Design and Development of Chromatic Dispersion Characterization System<br>*I. S. Amiri, M. Ghasemi and H. Ahmad* | **1** |
| **Chapter 2** | Analysis of the Results of the Modulation Phase Shift Method to Measure Chromatic Dispersion<br>*I. S. Amiri, M. Ghasemi and H. Ahmad* | **17** |
| **Chapter 3** | A Review of Optical Waveguide Sensors Using Ion-Exchange Technology<br>*I. S. Amiri, M. M. Ariannejad and H. Ahmad* | **33** |
| **Chapter 4** | Silver Nano Particle Ion-Exchanged Glass Waveguide Technology<br>*I. S. Amiri, M. M. Ariannejad and H. Ahmad* | **51** |
| **Chapter 5** | SysML: The Dialect for Model-Based Systems Engineering<br>*Ana Luísa Ramos and José Vasconcelos Ferreira* | **63** |
| **Chapter 6** | Complex System Analysis for Engineering of Systemic Failures<br>*Polinpapilinho F. Katina and Behnido Y. Calida* | **105** |
| **Index** | | **133** |

# PREFACE

This book discusses the concepts, tools and applications of systems engineering. Chapter One presents the implementation of automated measuring system for measuring chromatic dispersion (CD). Chapter Two discusses the results regarding the optimization of phase based EOM (electro optical modulator) as the most important component in CD measurement setup, then investigates the results of CD measurement characterizations based on different lengths of fiber, different wavelengths and different RF frequencies. Chapter Three provides a review on optical waveguide sensor using ion-exchange technology. Chapter Four examines silver nano particle ion-exchanged glass waveguide technology. Chapter Five highlights the major advantages of using graphical modelling languages in a Model-Based Systems Engineering (MBSE) context and describes the main features of SysML, the modeling language that the authors believe to be "the dialect" for MBSE. Chapter Six provides a complex system analysis for engineering of systemic failures.

Chapter 1 – This work presents the implementation of automated measuring system for measuring CD by using modulation phase shift method over long haul of optical single mode fibre. All forms of dispersion degenerate the modulation-phase relationships of light wave signals, decreasing information-carrying capacity through pulse-broadening in digital networks and distortion in analog system. Hence, designing system for measuring chromatic dispersion (CD) might be essential for estimation the performance of optical transceiver systems. Thus, the modulation phase shift (MPS) as one popular technique implemented for automatic measuring CD with the help of LabVIEW software, and in conjunction with programmable tunable laser (TLS) to provide the continuous optical wave at required wavelength, an electro optical modulator (EOM) for modulation of (radio frequency) RF signal on optical carrier and programmable digital oscilloscope (DOSC) for measuring phase difference between transmitted RF signal over long haul fiber and received RF signal at receiver. The working principles of the systems are comprehensively elucidated in this work. Additionally, the system components of the designs are identified and the systems are thoroughly characterized.

Chapter 2 - This chapter follows the proposed optical setup for measuring the chromatic dispersion (CD) of the long haul fiber through phase modulation difference method and will present the experimental results. This chapter, firstly discuss the results regarding the optimization of phase based EOM (electro optical modulator) as the most important component in CD measurement setup, then investigates the results of CD measurement characterizations based on different lengths of fiber, different wavelengths and different RF frequencies.

Chapter 3 - Optical interconnects are expected to complement or even replace electrical interconnects for shorter and shorter distances, as they allow for large data transmission rates as well as very high link densities. An attractive way to implement optical interconnects in modern multiprocessor systems is to integrate an optical waveguide layer into conventional printed circuit boards. A technology based on ion-exchange in commercially available glass substrates developed, as a fabrication technology for optical waveguides in the context of optical interconnects. Glass has unprecedented properties that make it the material of choice for many optical applications. Silver ion-exchange in glass reveals to be a highly promising method for the fabrication of low loss optical waveguides, especially also in the context of optical interconnects, because it is a scalable, industrially compatible and low-cost technology.

Chapter 4 - Ion exchange in glass and key developments in the first years of research are briefly described. Ion exchange in glass is a well-established method for fabrication of passive and active integrated photonic devices. An overview of fabrication, characterization and modeling of waveguides is given and the most important waveguide devices and their applications are discussed. Ion exchanged waveguide technology has served as an available platform for studies of general waveguide properties, integrated optics structures and devices, as well as applications. It is also a commercial fabrication technology for both passive and active waveguide components.

The generation of nano sized metal particles in amorphous matrix, the changes of the optical properties and the intensity of the reflective index were analyzed based on Surface Plasmon Resonance (SPR). Surface plasmons, particularly localized surface plasmons in the case of metal nanoparticles, explain the properties from the viewpoint of physical phenomenon and present the changes of the electromagnetic field due to the presence of metal particles.

Chapter 5 - A successful Model-Based Systems Engineering (MBSE) environment aims to construct a model of a system that can be transformed into the "real thing" through a given MBSE methodology. Modelling is critical to achieve the overall "maximum" performance of the system through a cooperative, integrative, adaptable and interoperable environment. So, the main artifact of a MBSE methodology is the System Model.

Model the system (or Design the system) is a central piece of the Systems Engineering' process and results in a complete low-level description of the system-of-interest that will guide the successful development and deployment of that system. This success is measured by the fulfilment of the stakeholders' expectations and, consequently, by their satisfaction.

The effective utilization of graphical modelling languages able to support collaborative development environments and successful stakeholders' communication and interactions constitutes a major issue when developing the System Model. SysML (Systems Modeling Language) is a UML profile or dialect for Systems Engineering and is currently the state-of-the-art modeling language for MBSE enhancing the development of successful systems in collaborative environments.

The chapter highlights the major advantages of using graphical modelling languages in a MBSE context and describes the main features of SysML, the modeling language that the authors believe to be "the dialect" for MBSE. This description is illustrated with original SysML diagrams related to the human body (probably, one of the most complex systems in our world).

Chapter 6 - At the very fundamental level, the utility of engineering approaches is to offer greater understanding of a given phenomenon so as to enable problem solving in real world situations. However, all engineering approaches are not of equivalent value – there are different benefits and limitations associated with each methodological approach for engineering. This chapter examines the nature of differences. First, a high level classification of methodology approaches is provided. The aim is this section to 'bin' different engineering methods based on how one obtains and communicates knowledge (i.e., underlying philosophy). Second, the type of failures that must be addressed by systems-based approaches are examined. This section seeks to provide a context-independent set of system failure paradigms that can be used to deal with different failures in complex systems. A context-independent conceptualization of systemic failures would include formulating an acceptable level of abstraction of the system. This ensures that underlying worldviews are made explicit and appropriately represented to reveal system pathologies. To help address this, a systems theory-based approach using system pathologies and the M-Path Method™ is introduced. This provides a framework through which perceived systems of interests can be analyzed to identify systemic failures. The chapter concludes with practical insights for how to effectively deal with complex system failures – including recent tools.

In: Systems Engineering
Editor: Marlene Hopkins
ISBN: 978-1-63485-752-9


*Chapter 1*

# DESIGN AND DEVELOPMENT OF CHROMATIC DISPERSION CHARACTERIZATION SYSTEM

***I. S. Amiri, M. Ghasemi and H. Ahmad***
Photonics Research Centre,
University of Malaya, Kuala Lumpur, Malaysia

## ABSTRACT

This work presents the implementation of automated measuring system for measuring CD by using modulation phase shift method over long haul of optical single mode fibre. All forms of dispersion degenerate the modulation-phase relationships of light wave signals, decreasing information-carrying capacity through pulse-broadening in digital networks and distortion in analog system. Hence, designing system for measuring chromatic dispersion (CD) might be essential for estimation the performance of optical transceiver systems. Thus, the modulation phase shift (MPS) as one popular technique implemented for automatic measuring CD with the help of LabVIEW software, and in conjunction with programmable tunable laser (TLS) to provide the continuous optical wave at required wavelength, an electro optical modulator (EOM) for modulation of (radio frequency) RF signal on optical carrier and programmable digital oscilloscope (DOSC) for measuring phase difference between transmitted RF signal over long haul fiber and received RF signal at receiver. The working principles of the systems are comprehensively elucidated in this work. Additionally, the system components of the designs are identified and the systems are thoroughly characterized.

**Keywords:** chromatic dispersion (CD), modulation phase shift, tunable laser (TLS), RF signal, programmable digital oscilloscope (DOSC)

# 1. INTRODUCTION

The broadening of light pulses, also known as dispersion, is an important factor degenerating the quality of signal transmission over optical fibers. This phenomenon is an outcome of the physical properties of the transmission medium. Single-mode fibers which are used abundantly in high-speed optical networks, are encounter the different scale of pulse broadening due to CD at certain wavelength as well as Polarization Mode Dispersion (PMD) which led to pulse broadening as a result of polarization [1-2]. The distorted pulse as result of broadening will cause bits exceed their intended time slots and overlap with their adjacent bits (Figure 1.1) [3]. This problem at receiver end may then create the problem of correct detection of interpreting adjacent bits, as a consequence increasing the Bit Error Rate. In any communication system keeping the transmission quality at standard level can be achieved by minimizing the dispersion [4]. The maximum amount of time dispersion must be limited to a small proportion of the signal bit rate, typically 10% of the bit time [5]. Before expansion the debate about dispersion, understanding the concept of refractive index is fundamentally important. This concept can be explained simply by measuring the speed of light at different medium. Basically, light travels at slower speed than in vacuum [6]. Light's speed become lesser within a medium than in vacuum. This speed usually determines by the refractive index of the medium and this is expected to be constant in response to different light's wavelength [7]. However, in real situation, the nature of the refractive index varies with variation of the wavelength and as one important contribution of this fact, light within the fiber travels at different speed when variation of wavelength is considered [8]. Basically, in an ideal case, the source of light must provide light with a unique wavelength, also the fully standardized laser sources are spectrally thin, but not enough thin to be defined as monochromatic. This clears that under the situation of quite narrow laser pulse, it contains at least several wavelength components which travels at different speeds, and finally the pulse would be broaden after travelling over the certain distance of the fiber [9-10]. The disadvantageous effect of CD result in slowing the speed of light component with property of slow wavelengths relevant in one pulse, and simultaneously allow them to travel with the faster spectral components of an adjacent pulse. This finally lead to different degree of intermixing between two very closed symbols which known as interference. This phenomenon is clearly shown in Figure 1.1. The CD of a fiber is usually defined in the form of ps/(nm*km), and is a factor to measure the imposed delay by this type of dispersion. Fundamentally, the chromatic dispersion means time spreading (in ps), for a source with a spectral width of 1 nm traveling on 1 km of the fiber. The quantity of CD depends on the fiber type, and based on this restricts the bit rate or the transmission distance for a good quality of service [10-14].

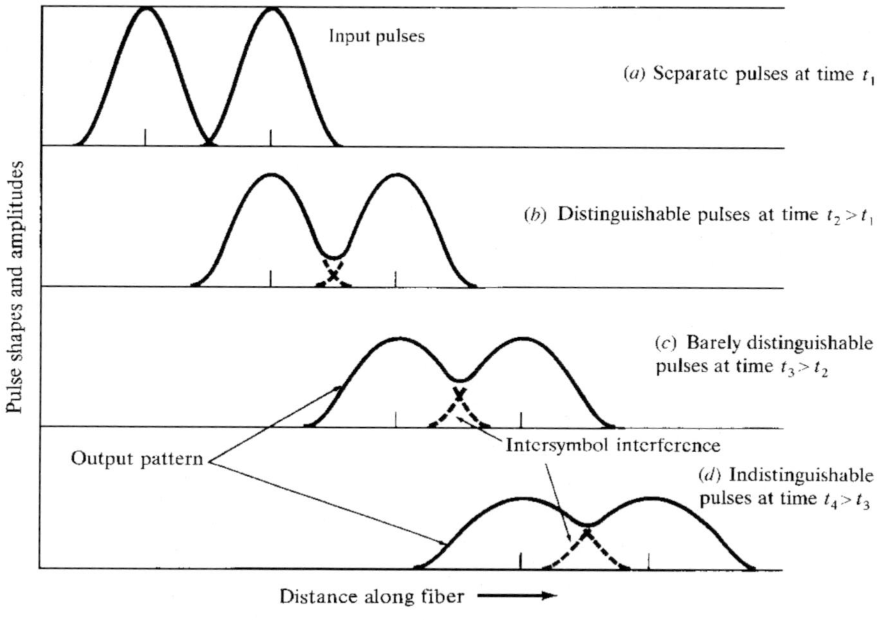


Figure 1.1. Broadening, attenuation and overlapping of two pulses while they pass through a fibre.

## 2. CD MEASUREMENT METHODS

Following the debate in previous section, it has been found three useful methods for measuring the CD of an optical fiber in industry. These are described by three TIA/EIA industry standards: the pulse-delay method

(FOTP-168 standard), the modulated phase-shift method (FOTP-169 standard), and the differential phase shift method (FOTP-175 standard). In all aforesaid methods, they firstly measure the time delay, in ps, respecting with the relevant wavelength then extract the approximate CD coefficient, in ps/(nm*km), from the slope of the achieved delay curve when the length of the link is enough long to observe and determine the slope. It must be noted, extraction of delay curve by the use each of aforementioned methods finally required to be fitted with a trace line since the acquired data of time delays is measured from discrete wavelength points. As such, the accuracy of the curve and CD coefficient relay on the quality and quantity of the measured data [3- 15].

In Figure 2.1 and 2.2, the schematic of phase-shift and methods are showing that these two methods are quite similar in concept. In both phase-shift methods, a sinusoidal modulated signal by optical carrier is sent to the input end of the fiber under test. The phase of received sinusoidal signal after de-multiplexer is measured at the output of the fiber and compared to the phase of transmitted signal which is known as reference signal. The comparison of transmitted and received signal must be conducted when both measured at the same frequency. In phase-shift method, we exploit the variation of wavelength of optical source rather than modulation of signal. Here, one optical source provides the reference signal which has a fixed wavelength, while the other modulated signal is tuned in wavelengths. In the differential phase-shift method, both signals are assumed to be tuned in wavelengths with equal span intervals. The received modulated signals then compared to a close reference signal which is tuned in a certain wavelength. It is important to make sure the gap between two wavelengths being constant in tuning process. Finally, the time delay of the long haul fiber is obtained from the phase-shift measurement, using the relationship between the delay (t), the phase (φ), and the modulation frequency (f) [16]:

$$\Delta t = \frac{\Delta \phi}{2\pi f}$$

The problem of phase-shift methods may explain by requirement of a link between transmitter and receiver stations for the fiber under the test. This link provides the phase of the reference signal for the receiver. Also, the process of tuning wavelength in differential phase shift method can be done either at the transmitter station by adjusting the tuneable laser, or at the receiver station by using broadband source and a wavelength tuneable filter. The differential method is known to be costly since the filter at the receiver must be spectrally enough thin (FWHM < 1 nm) to deliver the high measurement accuracy. One more disadvantage of this method is that the phase- shift need to use two signals with low interval wavelengths. Increasing the span of two adjacent wavelengths corresponding to two modulated signals will enhance the accuracy of the delay at selected wavelength point, but will reduce the number data points needed be traced for plotting the delay curve, hence, it decreases the accuracy of the measured CD coefficient. A trade-off must be agreed between the wavelength interval for the differential phase-shift calculation and number of points must be obtained for getting a fit the delay curve.

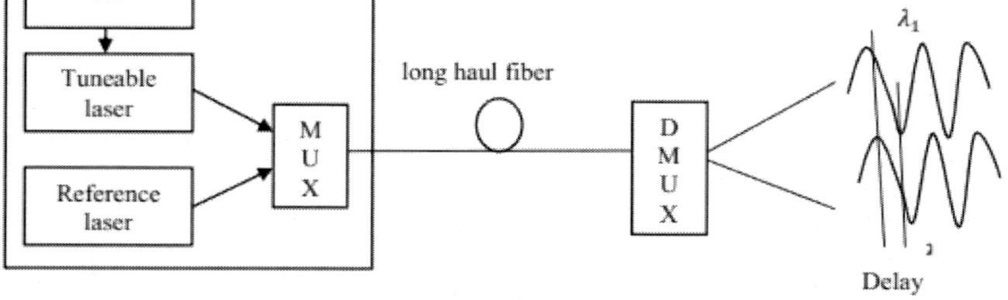


Figure 2.1. Fundamental setup for measuring phase shift.

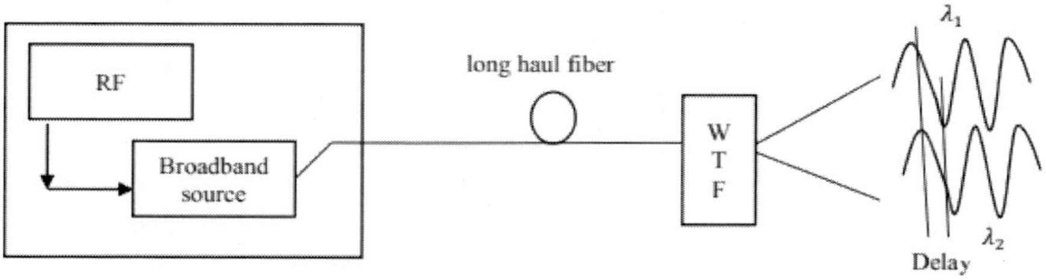


Figure 2.2. Fundamental setup for differential phase shift measuring using wavelength tuneable laser filter at the receiver.

The above discussion clearly shown that phase-shift methods are two-end of long haul fiber for CD measurement. However, the optical amplifier is also required when very long haul fiber (>100 km) is considered between transmitter and receiver. This method is trustable, since all the information required for the measurement are obtained by participating of two ends. Also, the wavelengths of modulated signals are also known, this enhance the efficiency of measurements by increasing the points of measurements and a better fit of the delay curve. Practically, the implementation of these setups are expensive and thus may not be suitable for commercialized applications. Thus in the next section the setup has been reconfigured in order to reduce the cost and the complexity of tuning wavelength at the receiver or transmitter by employing only one tuneable laser source combined with electro-optical modulator at transmitter end and only one light to RF wave convertor at the receiver end [17, 23].

# 3. Experimental Setup for Measuring Dispersion Using Phase Shift Method over Long Haul of Fibre

The specification of the electrical and optical components used in the CD measurement setup presented in Figure1 will be discussed in this section. This part is important step before configuring the setup and running it. As it is evident from Figure 1, the setup comprises of tuneable laser, digital oscilloscope, external modulator, single mode fibre, polarization controller, laser diode, power supplier and function generator. From the schematic of the setup shown in Figure 3.1, it can be seen the phase shift of RF signal performed as optical signal, simply by replacement of wavelength tuneable laser filter at the receiver with a light wave converter. Thus the role of tuneable is providing an optical carrier for electro optical modulator. The phase shift can be measured by a simple digital oscilloscope at the receiver end as well. Whole the progress of CD measurement either at transmitter/ receiver can control by the controller defined in LabVIEW software. Figure 3.1 illustrates that the LabVIEW controller starts the process of measurement by tuning the laser source at transmission end. The laser beam at selected wavelength inject to electro optical modulator which adjusted to operate at high efficient point by applying a suitable DC voltage. Meanwhile, the frequency of the RF is monitored by LabVIEW controller in channel 1. After receiving signal and demodulation back into RF signal at the receiver end, LabVIEW controller retrieve the phase of received RF signal from channel 2 and compute the phase difference between these two channels. The modulated RF signal on optical carrier inject again to long haul fiber. The next subsections will mainly focus on the important features of the electric and optical components which taken from their data sheets. The extraction of phase difference between two channels continues on till whole the selected wavelength range provide enough data for creation of dispersion curve.

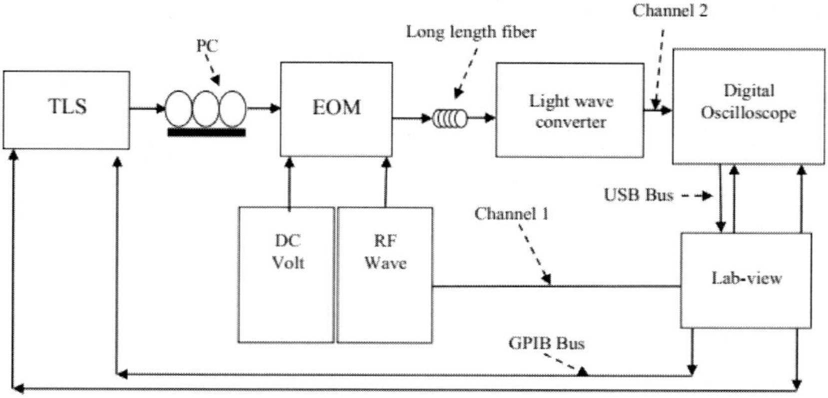


Figure 3.1. Schematic setup for measuring CD of long length fibre (several kilometres distance) using phase shift method.

## 3.1. Tuneable Laser Source (TLS)

The tuneable laser model AQ8201-13 has been selected for producing and process of sweeping laser within the range of 1300 nm to 1500 nm. The block diagram of this device as to the manufacturer data sheet is shown in Figure 3.2.

As it is evident from Figure 3.2, the system initially starts by controller which has the task of observation the situation of the driver and temperature of the optical module when the key enter trigged by the user or software. Next, the controller with respect of the demanded wavelength and power by the user/ software through the signals received from the frame activate the module interface for driving the optical module. Meanwhile, the temperature is the key factor for both controller unit and optical module during engaging the driver running the pulse motor. It must be noticed the wavelength can be set by variation of the pulse motor. After setting the power as to selected wavelength then it will launch from optical module unit to optical connector which is a simple interface to convey the light to outside the tuneable laser. The main features of this device is listed in Table 3.1. In this table, RIN is the ratio of the carrier power and noise power per unit of bandwidth. The optical connector is selected as FC/PC which is a typical ultra-polished FC (fiber optic connector) with very low insertion loss.

**Table 3.1. AQ8201-13 tuneable laser main features**

| | |
|---|---|
| Wavelength range | 1460-1580nm |
| Wavelength resolution | 10pm |
| Optical output level in the range 1520-1580 | +10dBm |
| Operating temperature | 23±5° |
| RIN | -145 dB/Hz |
| Optical connector | FC/PC |

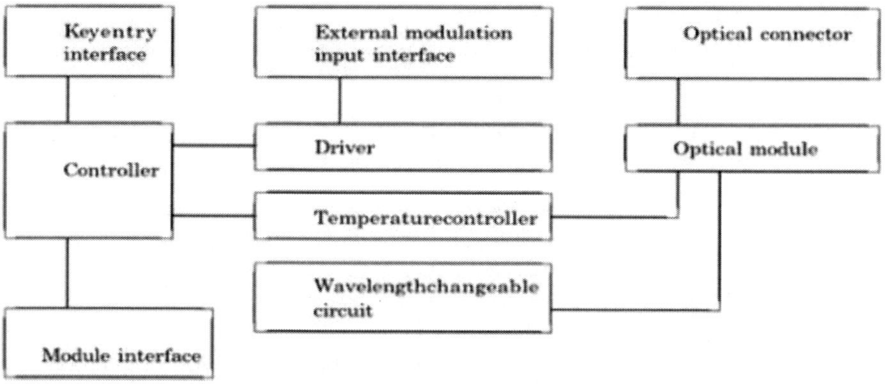


Figure 3.2. Shows the connectivity and functionality of each part in tuneable laser module AQ8201-13.

## 3.2. Polarization Controller (PC)

The main purpose of using PC for this experiment is related with involving electro optical modulator within the setup. It is required to adjust the polarization state of laser in the fibre before it enters to the external modulator. The birefringence phenomenon in this case helps for controlling polarization and can be applied by coiling fibre. The induced retardation is corresponding with two factors, first the length of fibre and the second inverse of bending radius. The FPC 560 is one of the popular polarization controllers and classified as "BAT EAR" controllers because it uses some holders at specific diameters and adjustable for certain number of fibre turns in order to induce the retardation.

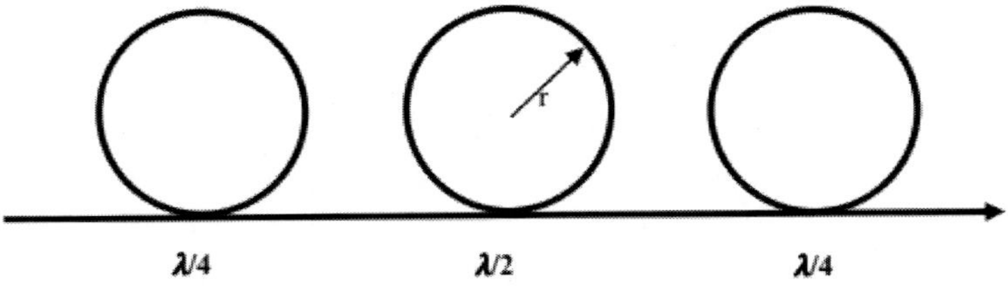


Figure 3.3. "*BAT EAR*" polarization controller.

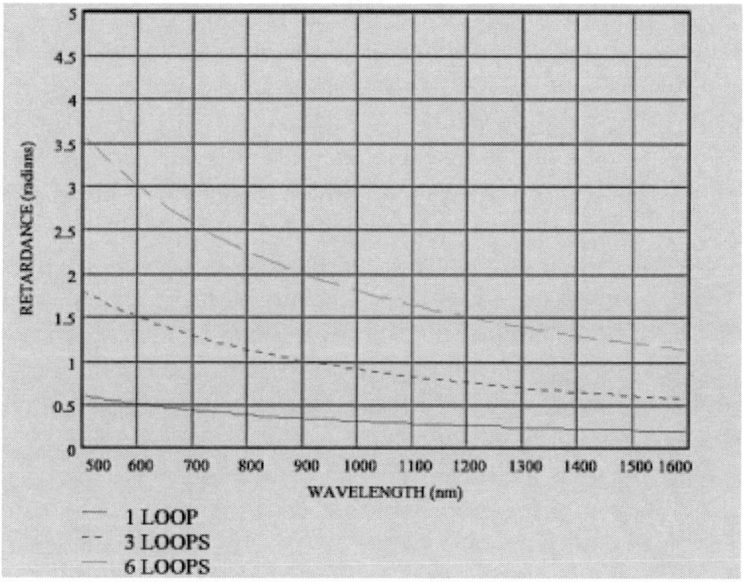


Figure 3.4. Retardation vs. Wavelength for 1, 3, and 6 Fiber Loops Per Paddle. The Fiber Clad Diameter is 80μm.

Each paddle in Figure3.3 can make effective retardation corresponding to the number of fibre turns at specific diameter (if the paddle diameter is too small the bending loss could be happened). For FPC560 the first and the second and the third paddles make $\lambda/4$, $\lambda/2$ and $\lambda/4$retardation, respectively. By changing the angle of first paddle the input state polarization state converts into linear polarization state and the third state would convert it into the arbitrary output polarization state at a fixed wavelength. Since the polarization state is function of wavelength, the high peak power can be achieved due to polarization rotation. Figure3.4 depicts how at lower wavelengths and higher number of loops retardation takes the greater values compare with higher wavelengths.

## 3.3. Lithium Niobate (LiNbO3) Electro-Optic Modulator (EOM)

The ferroelectric material lithium niobate (LiNbO3) is one of the most well-known optical modulator which is used extensively in optical devices. This modulator found quite suitable for its electro-optic and optical properties [18]. The two distinguished features of this device are large electro-optic effect and capability of high-speed response. One of the important optical property of this modulator is its transparency in infrared frequency regime; It also can fabricate into low-loss channel waveguide at very low cost fabricate through mechanism of diffusing the titanium. The different optical waveguide based on LiNbO3 with excellent degree of performance have been designed for the applications of terminal functions of external intensity modulators, phase modulators, and multi/de-multiplexers, as well as switch arrays for optical fiber network systems. In a special application of optical communication, the transparency of LiNbO3 devices are found very suitable for WDM systems since they can operating efficiently in the range of wide-wavelength infrared light as a single device [19]. It must be emphasized that the LiNbO3 external modulators have been designed in order to operate in high speed and long-distance optical fiber transmission systems. One of the reason to use this type of electro-optical modulator is because they can provide the data rate modulation exceeding 10 Gbits/sec combined with very low driving voltage. One advantage point which is consider in designating this type of external modulators is they have the property to reduce the dynamic laser wavelength chirping as factor restricting the span-rate system product completely. This feature mainly supported by the anti- dispersion characteristics of the use fiber in the structure of the modulator. Besides, LiNbO3 external modulators play as an important role in optical signal processors due to the ability of pure phase modulation in coherent systems. With the increasing in the bit rate of optical network systems [20], it found more difficult to increase the driving voltage after reaching saturation point as a result of the restriction caused by electrical instruments, in particular, electrical driving amplifiers. Therefore, keeping the driving voltage of an LiNbO3 modulator at minimum level in the communication systems with a broadband characteristic is an extremely important issue for realizing future high-speed optical transmission systems [21].

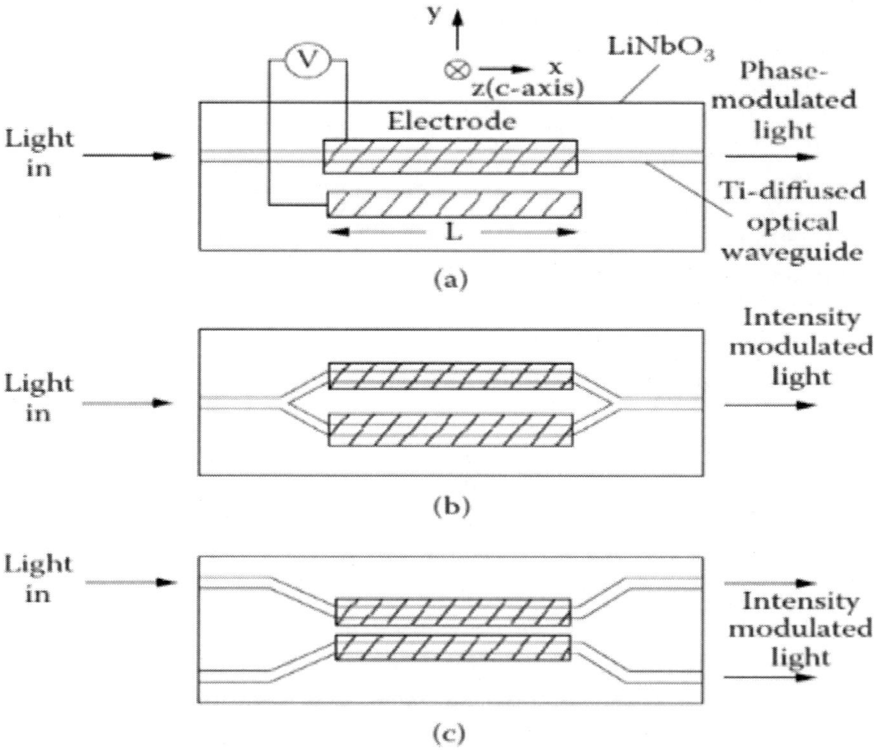


Figure 3.5. Z-view of typical LiNbO3 devices. (a) phase modulator, (b) intensity modulator using Mach-Zehnder interferometer, and (c) intensity modulator using directional coupler.

### *3.3.1. Different Modulation Schemes of $LiNbO_3$ Electro-optic Modulator Devices*

In all schemes presented in Figure 3.5 the channel of the waveguide is considered to be single-mode channel device. The z-view of the devices shows that the $LiNbO_3$ crystals is used as a substrate, and a low-refractive-index buffer layer of $SiO_2$ film is designed between the electrode and the $LiNbO_3$ substrate to prevent the further absorption of the light by the metal. The external modulators which is used in the setup presented in Figure 3.1a, is a phase modulator device based on $LiNbO_3$ material. It is evident, the phase modulated optical guide between electrodes and $LiNbO_3$ is not purely made from $SiO_2$ layer and titanium has been diffused in the body of the channel to enhance performance of the optical channel. In Figure 3.2b and 3.2c, the absence of optical wavelength chirping in modulated signals is essential in high-speed and long-distance optical transmission systems [22].

**Table 3.2. Importance operating conditions of LiNbO3 phase modulator**

| | |
|---|---|
| Operating Wavelength | 1525nm to 1605nm |
| Optical Insertion Loss (2) | 5 dB |
| RF $V\pi$ Voltage | 3.8-5 Volts@1kHz |
| Bias $V\pi$ Voltage | 5.5-6Volts@1kHz |
| Optical Return Loss | 45dB |
| DC Optical Extinction Ratio | 20-24dB |
| Chirp, Alpha parameter | -0.1 to 0.1 |

## 3.4. Programmable Signal Generator

The RF source for the external modulator is produced by PM5191function generator. It has capability to produce different waves like sine and square form. It also work within the frequency range between 0.1 mHz to 2 MHz. Accuracy and stability are the essential factors for produced sine wave that fulfils with this synthesizer/function generator. Other important feature of PM5191 is being remote programmability using GPIB port. All front panel functions can be programmed remotely and all settings and status data can be called and controlled through the remote controller. Some major features about the function generator are as follow:

- Sine wave 0.1mHz-2.147MHz
- Square wave 0.1mHz-2.147MHz
- Phase noise < -80 dBc/Hz
- Signal to noise ratio (SNR) ≥ 55dBc
- Long term drift < 0.3ppm with in 7 hours
- Max resolution 0.1 mHz
- Voltage peak to peak open circuit (for both sine wave and square wave) 0-30 Volts
- Rise and fall time for square wave <35ns
- Duty cycle 50%

## 3.5. Optical-to Electrical Converter (O/E)

In many optical application sometimes it is required to convert the optical signal into very low frequency signal (RF). For this purpose, Agilent 11982A is utilized by the mean of PIN photo detector to convert optic signal to electric form. For optical application the PIN photo detector can support bandwidths of the order of tens of gigahertz. After conversion process, the produced electrical signal needs to be amplified by a very low noise amplifier due to highly attenuation during conversion. The amplified electrical signal is suitable to use with measuring instrument such as digital oscilloscope or spectrum analyser. In frequency domain optical characteristics of laser beam such as Intensity modulation, distortion, and laser intensity noise can be displayed by spectrum analyser and the effect of laser modulation is measurable. Other important capability of the 11982A is changing display oscilloscope unit to watts unit in order to measure optical power. For this purpose it is enough to enter the reciprocal of the responsivity in oscilloscope's probe attenuation field. Considerable features of the O/E 11982A are as follow:

- Wavelength Range 1200 nm to 1600 nm
- Bandwidth optical dc to 15 GHz
- Bandwidth electrical dc to 11 GHz
- Input return loss 23dB
- Conversion gain(accuracy of provided value) ±20%
- Maximum safe optical input power(average) 10mW
- Temperature range 0°C to +55°C

## 3.6. Programmable Digital Signal Oscilloscope

The "DSO1022a," from Agilent.co has been used to analyse the RF signal after conversion by O/E at receiver end (Figure 3.1). This device supports two input RF channels. Automatic measurement function in this model of oscilloscope provides many facilities for measuring amplitude and time components of the received signal. LabVIEW controller in Figure 3.1, has sent a request continually to DSO1022a through GPIB cable link at equal interval of time to accurately measure phase difference of received and transmitted RF signals.

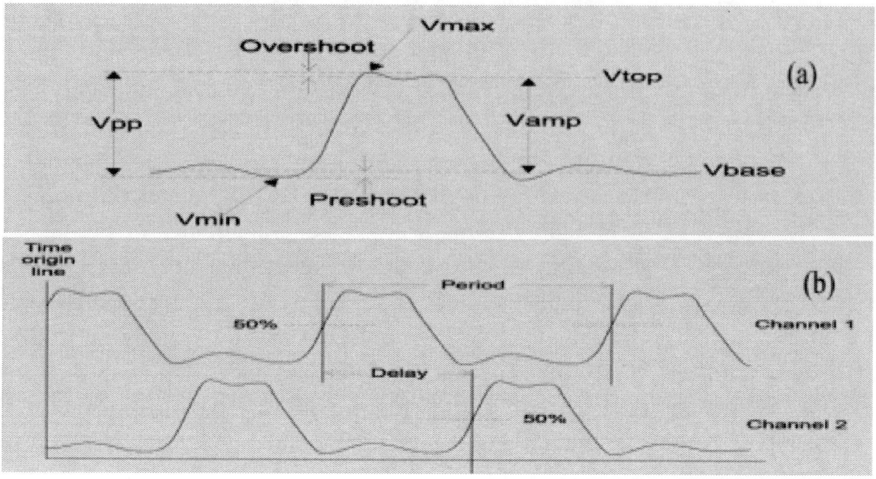


Figure 3.6. Measurable parameters of voltage (a) and time (b) components of the RF signal.

# REFERENCES


[1] Alwayn, Vivek. *Optical network design and implementation.* Cisco Press, 2004.

[2] Gumaste, Ashwin, and Tony Antony. DWDM network designs and engineering solutions. Cisco Press, 2003.

[3] Keiser, Gerd. "Optical fiber communication." NY: McGraw-Hill (2000).

[4] Rosenfeldt, Harald. Measurement and Compensation of Polarization Mode Dispersion in Optical Communication Systems. Cuvillier Verlag, 2006.

[5] Lecoy, Pierre. Fibre-Optic Communications. John Wiley & Sons, 2013.

[6] Band, Yehuda B. Light and matter: electromagnetism, optics, spectroscopy and lasers. Vol. 1. John Wiley & Sons, 2006.

[7] SINGH, LAKHMIR, and MANJIT KAUR. Science for Ninth Class (Part-1) Physics. RSM Press.

[8] Oliviero, Andrew, and Bill Woodward. Cabling: the complete guide to copper and fiber-optic networking. John Wiley & Sons, 2014.

[9] Hecht, Jeff. Understanding fiber optics. Jeff Hecht, 2015.

[10] Yao, Jianquan, and Yuyue Wang. Nonlinear Optics and Solid-State Lasers: Advanced Concepts, Tuning-Fundamentals and Applications. Vol. 164. Springer Science & Business Media, 2012.

[11] Buck, John A. Fundamentals of optical fibers. John Wiley & Sons, 2004.
[12] Ghatak, Ajoy, and K. Thyagarajan. An introduction to fiber optics. Cambridge university press, 1998.
[13] Barnoski, Michael, ed. Fundamentals of optical fiber communications. Elsevier, 2012.
[14] Agrawal, Govind P. Nonlinear fiber optics. Academic press, 2007.
[15] Laferrière, J., G. Lietaert, R. Taws, and S. Wolszczok. Reference guide to fiber optic testing. JDS Uniphase Corporation, 2007.
[16] Chan, Calvin CK, ed. Optical performance monitoring: advanced techniques for next-generation photonic networks. Academic Press, 2010.
[17] Forestieri, Enrico, ed. Optical Communication Theory and Techniques. Springer Science & Business Media, 2004.
[18] Sugiyama, Masaki, Masaharu Doi, Shinji Taniguchi, Tadao Nakazawa, and Hiroshi Onaka. "Driver-less 40 Gb/s LiNbO3 modulator with sub-1 V drive voltage." In *Optical Fiber Communication Conference*, p. FB6. Optical Society of America, 2002.
[19] Mohammed, Abd El-Naser A., and A. Mohammed. "Metawe'e, Ahmed Nabih Zaki Rashed, and Mahmoud M." Eid,"Important Role of Optical Add Drop Multiplexers (OADMs) With Different Multiplexing Techniques in Optical Communication Networks," *IJCIIS International Journal of Computational Intelligence and Information Security* 1, no. 1 (2010): 72-85.
[20] Courjal, N., A. Martinez, and H. Porte. "LiNbO3 Mach–Zehnder Modulator with chirp adjusted by domain inversion." *Proceeding Integrated Photonic Research* (2002): p-IFA4.
[21] Shimotsu, S., S. Oikawa, T. Saitou, N. Mitsugi, K. Kubodera, T. Kawanishi, and M. Izutsu. "Single side-band modulation performance of a LiNbO 3 integrated modulator consisting of four-phase modulator waveguides."*Photonics Technology Letters, IEEE* 13, no. 4 (2001): 364-366.
[22] Mohammed, Abd El-Naser A., Mohamed Metwae'e, Ahmed Nabih Zaki Rashed, and Amira IM Bendary. "Recent progress of LiNbO3 based electrooptic modulators with non return to zero (NRZ) coding in high speed photonic networks." *International Journal* (2011).
[23] Hui, Rongqing, and Maurice O'Sullivan. Fiber optic measurement techniques. Academic Press, 2009.

In: Systems Engineering
Editor: Marlene Hopkins
ISBN: 978-1-63485-752-9


*Chapter 2*

# ANALYSIS OF THE RESULTS OF THE MODULATION PHASE SHIFT METHOD TO MEASURE CHROMATIC DISPERSION

***I. S. Amiri, M. Ghasemi and H. Ahmad***
Photonics Research Centre, University of Malaya,
Kuala Lumpur, Malaysia

## ABSTRACT

This chapter follows the proposed optical setup for measuring the chromatic dispersion (CD) of the long haul fiber through phase modulation difference method and will present the experimental results. This chapter, firstly discuss the results regarding the optimization of phase based EOM (electro optical modulator) as the most important component in CD measurement setup, then investigates the results of CD measurement characterizations based on different lengths of fiber, different wavelengths and different RF frequencies.

**Keywords:** chromatic dispersion, electro optical modulator, RF frequencies

# 1. Characterization of EOM Modulator to Find the Best Operating Point

EOM modulator requires the biasing conditions before perform as a modulator in any optical setup. Since the phase modulator supposed to modulate the RF signal over an optical carrier, it must be first set in normal operating point electrically, then the RF signal applied for the purpose of modulation [1-9]. It must be noted at the right operating point, the external modulator works ideally and the non-distorted electrical signal would modulate on optical beam. DC biasing of the EOM is comprising the applying DC voltage to the electrodes of EOM and simultaneously measuring the variation optical power coming out from the modulator. It is expected the following sinusoidal figure is observed after variation of DC voltage from 0 to 10 volt.

In Figure 1, the recommended region for normal modulation of a complete peak to peak RF signal ($V_{p\text{-}p}$) highlighted with closed black solid-line loop, and beyond this region the signal most likely encounters the nonlinearity effect of EOM modulator. The symbol $V_\pi$ in Figure 1allows the RF signal to vary with in the boundary highlighted with orange solid line and above of these line the phase of RF signal become negative [10-16]. The characterization of different EOM phase modulators as shown in Table 1 indicates that they can successfully modulate a RF signal at different condition of DC bias voltage. Also the results at each time measurement explains the results would not be consistent essentially.

The results of characterization also emphasized on the importance of $V_b$parameter. The RF signal offer 180° shifts before and after the minimum point of the transmission transfer curve.

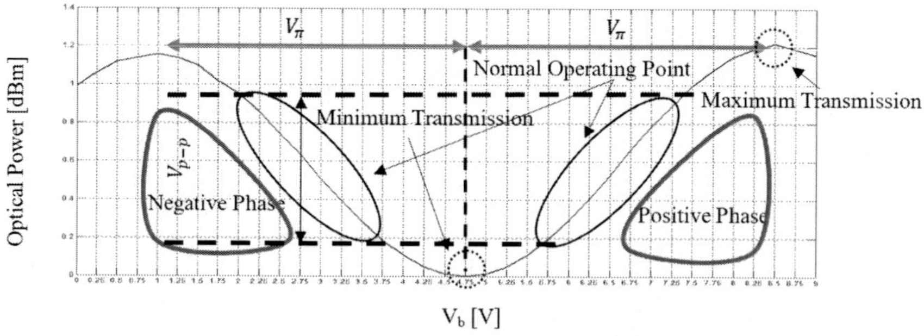


Figure 2.1. EOM transmission transfer curves versus DC bias voltage.

**Table 2.1. Characterization of EOM modulators**

| Negative phase area | | | | |
|---|---|---|---|---|
| Model No. | Test number | $V_b$(volt) | $V_\pi$(volt) | $V_p$(volt) |
| EOM-7F6E1900 | 1 | 2-4 | 1.5-5.5 | 2 |
| EOM-7F6E1900 | 2 | 2-4 | 1-5.5 | 2 |
| EOM-7F6E1900 | 3 | 2-4 | 1.5-5.5 | 2 |
| EOM-7F6E1900 | 4 | 2.5-4 | 1.25-5.5 | 1.5 |
| Positive phase area | | | | |
| Model No. | Test number | $V_b$(volt) | $V_\pi$(volt) | $V_p$(volt) |
| EOM-7F6E1900 | 1 | 6.5-8.5 | 5.5-9 | 2 |
| EOM-7F6E1900 | 2 | 6.5-8.5 | 5.5-9 | 2 |
| EOM-7F6E1900 | 3 | 7-9 | 5.5-9.5 | 2 |
| EOM-7F6E1900 | 4 | 7-9 | 5.5-9.5 | 2 |
| Negative phase area | | | | |
| Model No. | Test number | $V_b$(volt) | $V_\pi$(volt) | $V_p$(vol) |
| EOM-7F6E1A00 | 1 | 3-4.5 | 2-6 | 1.5 |
| EOM-7F6E1A00 | 2 | 3-4.5 | 1.5-6 | 1.5 |
| EOM-7F6E1A00 | 3 | 2.5-4.5 | 1.5-5.5 | 2 |
| EOM-7F6E1A00 | 4 | 2.5-4.5 | 1-6 | 2 |
| Positive phase area | | | | |
| Model No. | Test number | $V_b$(volt) | $V_\pi$(volt) | $V_p$(volt) |
| EOM-7F6E1A00 | 1 | 7-8.5 | 6-10 | 1.5 |
| EOM-7F6E1A00 | 2 | 7-8.5 | 6-9.5 | 1.5 |
| EOM-7F6E1A00 | 3 | 7-8.5 | 6-9.5 | 1.5 |
| EOM-7F6E1A00 | 4 | 7-8.5 | 6-9.5 | 1.5 |

In general, the EOM response is linear for small fluctuation of RF signal within the recommended regions. The transmission light power I(t) might be written in the form of Equation 1.1. From this equation it is evident that sin shape of transmission transfer curves which experimentally measured is compatible with the theory of power in electro optical modulator. Further, three basic components $V_b$, $V_\pi$ and $V_{p-p}$ must selected wisely as to their influence on output power of EOM device. In Equation 1, $I_0$ is the maximum transmitted light power and the $E(w)$ represents the frequency dependency modulation efficiency.

$$I(t) = \frac{I_0}{2}[1 + \cos\left(\pi \frac{V_b + E(w)V_p \cos(wt)}{V_\pi}\right) \quad (1)$$

The dc bias variable $V_b$ (Equation 2) is adjustable parameters and controls the operating point of the EOM. This parameter also is the key point subjecting to the shifting of transmission transfer curve when a DC bias voltage out of the permitted range applied to EOM device. In this situation, multiple operating point might be observed from transmission transfer curves (Figure 2).

The shifting phenomenon in transmission transfer curves of EOM causes the linear operation area become smaller and consequently the $V_{p-p}$ fluctuates in a very small region of recommended regions or normal operating point. The detection of high attenuated RF signal after travelling along a long haul fiber is difficult by the measurement device at receiver end. The characterization of EOM model No. 7F6E1900, in figures 2.4-2.13 Explains how the quality of the received RF signal after travelling along several meters fiber can vary as to increasing of $V_b$.

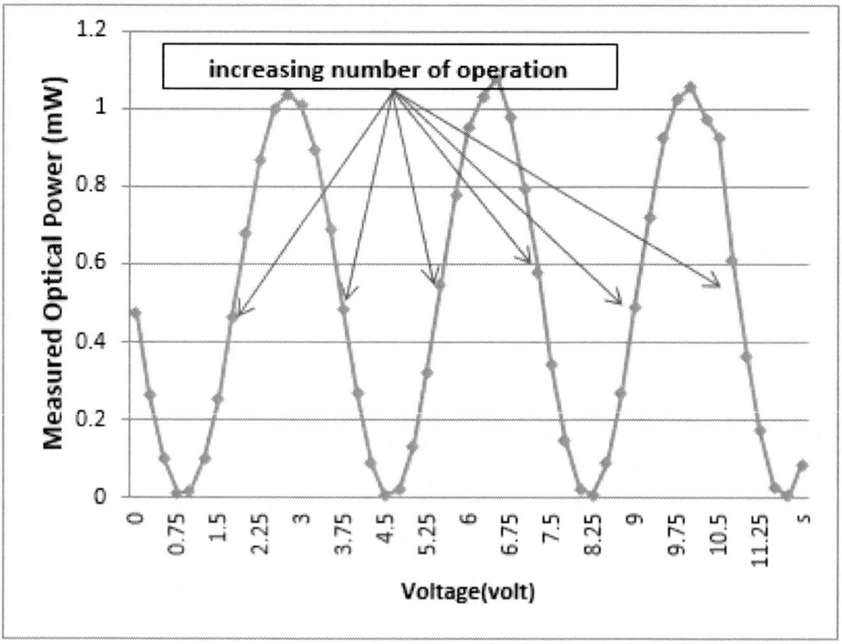


Figure 2.3. Shifting in transmission transfer curve as result of applying high amount of $V_b$.

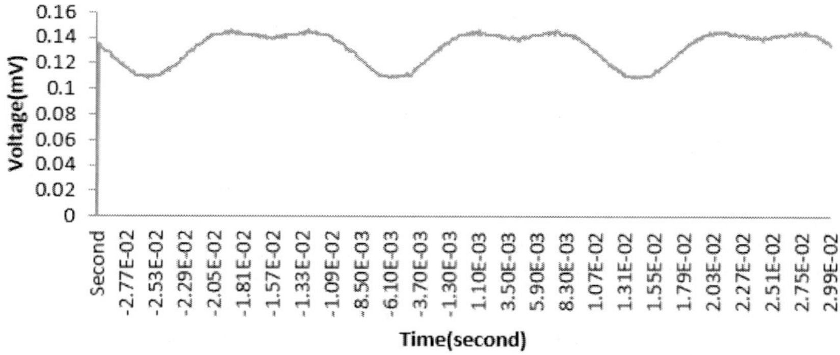


Figure 2.4. Measured RF signal at the receiver at $V_b = 2.5$ volt.

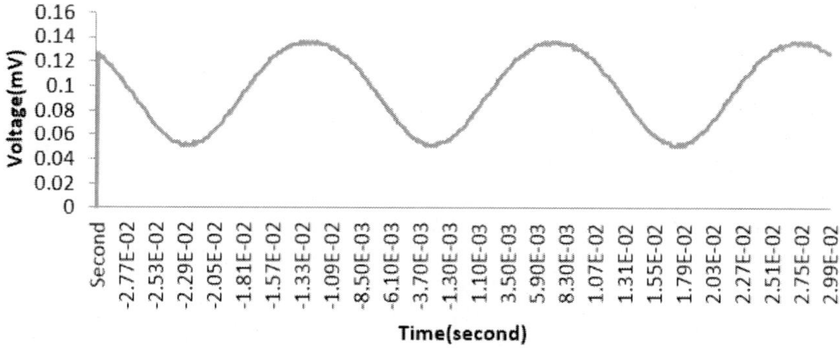


Figure 2.5. Measured RF signal at the receiver at $V_b = 3$ volt.

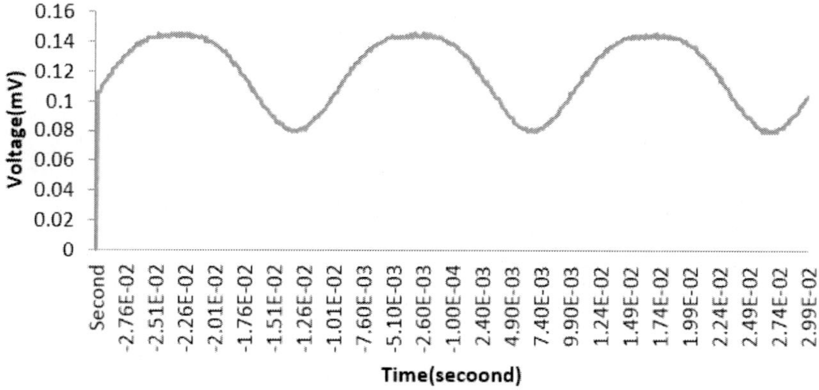


Figure 2.6. Measured RF signal at the receiver at $V_b = 3.5$ volt.

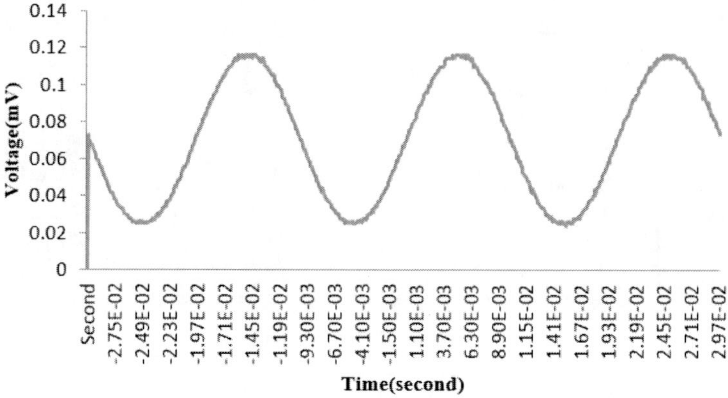


Figure 2.7. Measured RF signal at the receiver at $V_b$= 4 volt.

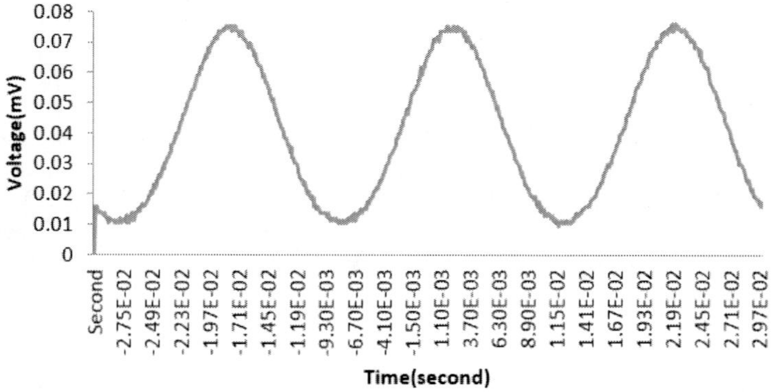


Figure 2.8. Measured RF signal at the receiver at $V_b$= 4.5 volt.

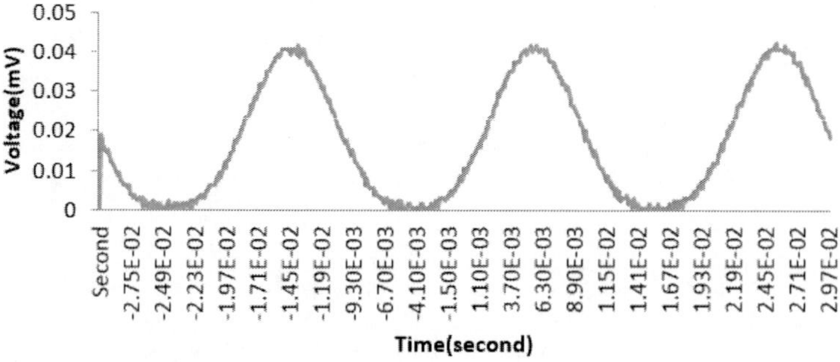


Figure 2.9. Measured RF signal at the receiver at $V_b$= 5 volt.

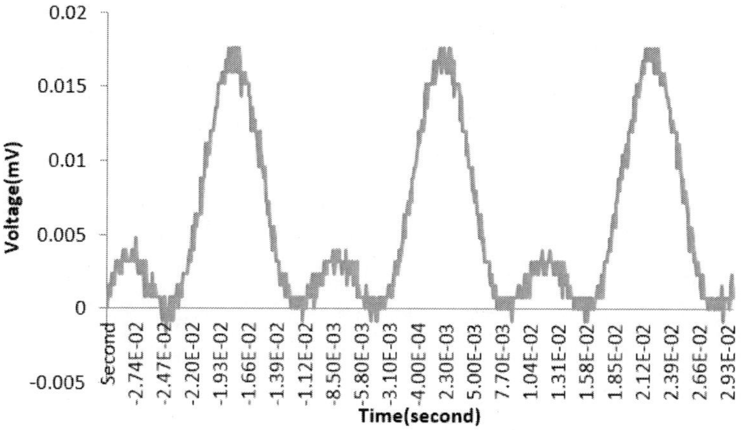


Figure 2.10. Measured RF signal at the receiver at $V_b$= 5.5 volt.

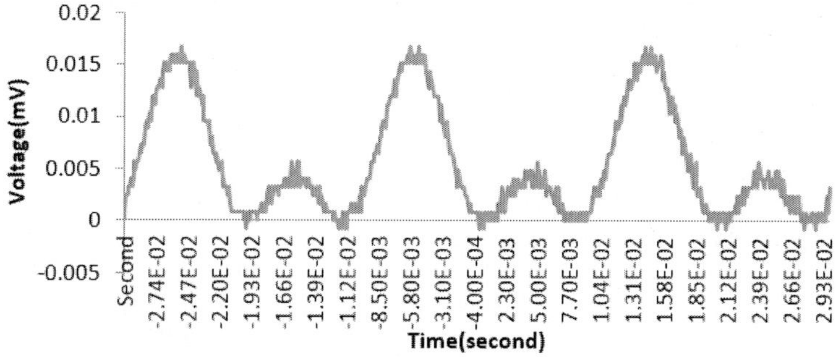


Figure 2.11. Measured RF signal at the receiver at $V_b$= 6 volt.

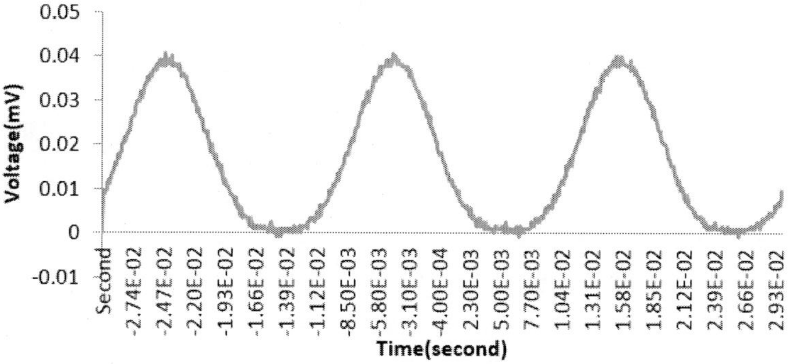


Figure 2.12. Measured RF signal at the receiver at $V_b$= 6.5 volt.

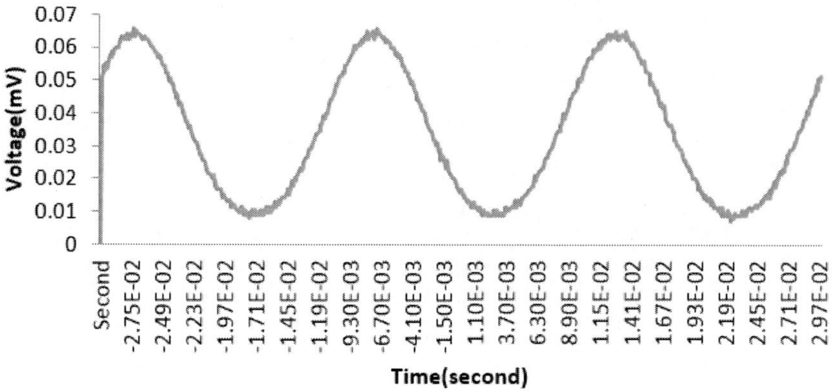


Figure 2.13. Measured RF signal at the receiver at $V_b$= 7 volt.

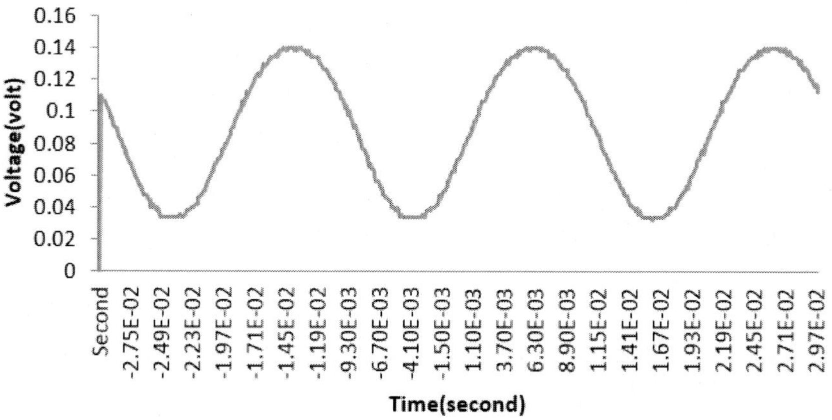


Figure 2.14. Measured RF signal at the receiver at $V_b$= 7 volt and $V_p$ = 1.96 volt.

The shape of the measured RF signals in figures 4-13 clearly illustrates the response of EOM to different values of $V_b$. Since the boundary of linear operation is not just limited to a specific amount of $V_b$, the highest similarity between the reference and measured RF wave can be used to select proper operation point. Besides the limitation posed by $V_b$, $V_{p-p}$ is limited to a certain range of fluctuate. Considering the bets operating point ($V_b$) from the Figure 4-13 is selected as 7 volt. The EOM model No. 7F6E1900 under the selected operation need to be characterized again to identify the best value of $V_{p-p}$ and before final evaluation of the setup for measuring the dispersion under long haul fiber.

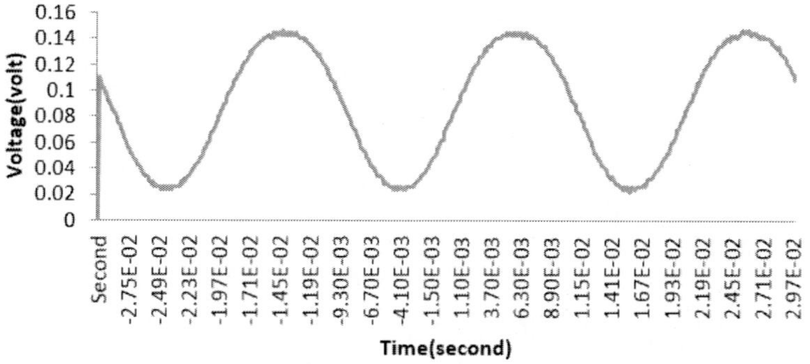


Figure 2.15. Measured RF signal at the receiver at $V_b$= 7 volt and $V_p$ = 2.64 volt.

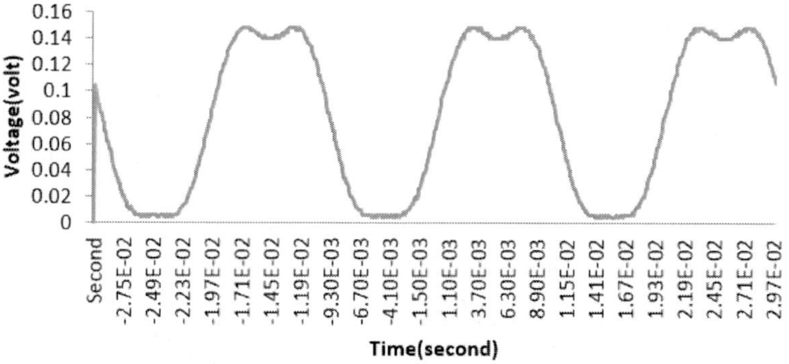


Figure 2.16. Measured RF signal at the receiver at $V_b$= 7 volt and $V_p$ = 4.16 volt.

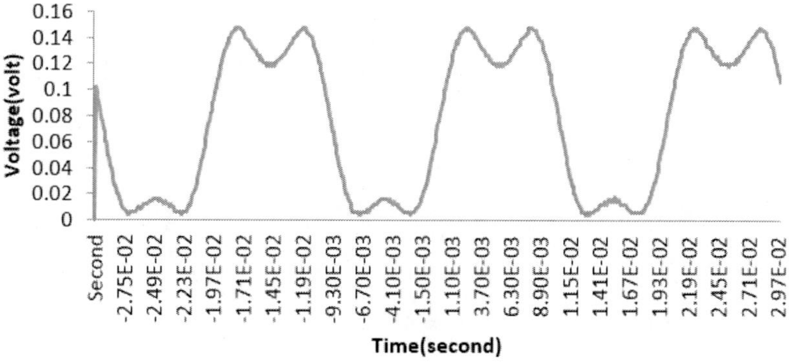


Figure 2.17. Measured RF signal at the receiver at $V_b$= 7 volt and $V_p$ = 5.28 volt.

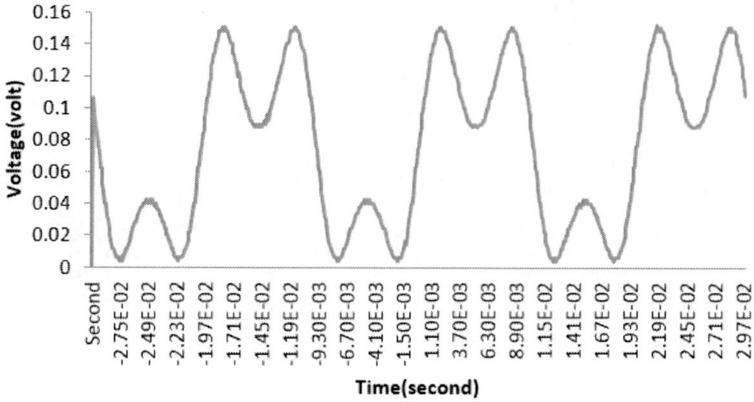


Figure 2.18. Measured RF signal at the receiver at $V_b$= 7 volt and $V_p$ = 6.4 volt.

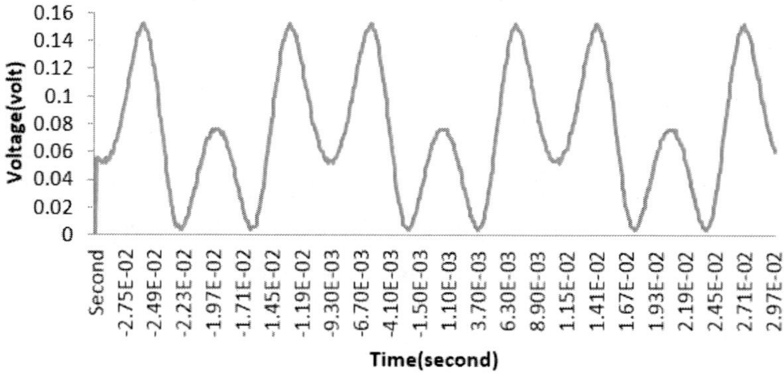


Figure 2.19. Measured RF signal at the receiver at $V_b$= 7 volt and $V_p$ = 7.52 volt.

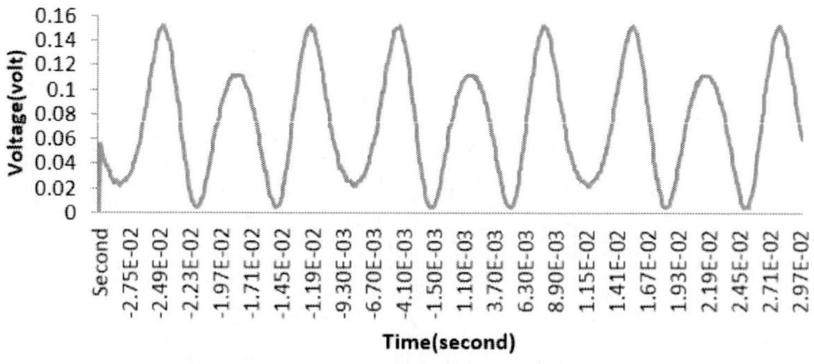


Figure 2.20. Measured RF signal at the receiver at $V_b$= 7 volt and $V_p$ = 8.64 volt.

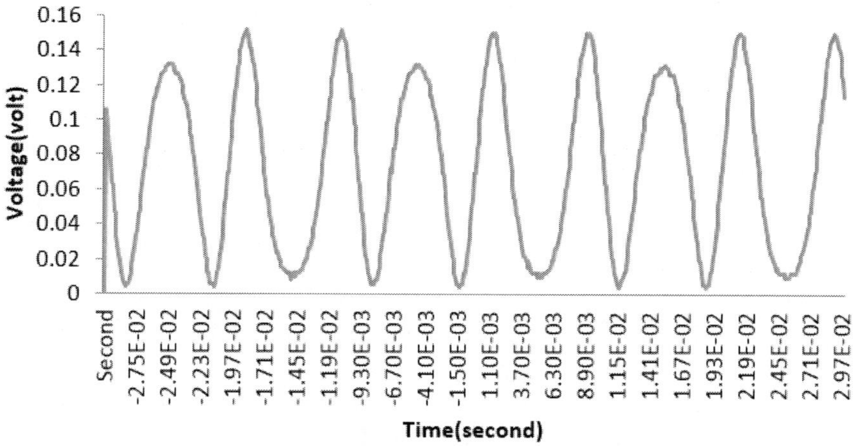


Figure 2.21. Measured RF signal at the receiver at $V_b$= 7 volt and $V_p$ = 9.36 volt.

The normal shape of signals from Figure 2.14 and 2.15 can also use to determine that the optimum operation point for EOM model No. 7F6E1900. Thereby, it can be selected as $1.96 < V_p < 2.64$ at $V_b$= 7. Besides, these two important points, $V_\pi$ or driving voltage of the EOM as shown in Equation 2 is under affection of more parameters such as wavelength and refractive index of LiNbO3[2].

$$V_\pi = \frac{\lambda G}{2n_e^3 r_{33} \Gamma L} = \frac{\lambda . V}{4.n(V).L} \qquad (2)$$

where $\lambda$ is the wavelength of applied laser beam, $n_e$ is refractive index of LiNbO3 and $r_{33}$is the electro optic coefficient constant $(3 \times 10^{-12} m/V)$ .The parameter V similar to $V_b$ is DC biasing voltage of the EOM. Parameter Γ is the overlap integral between optical and electrical (RF) fields. Furthermore, G is the free space gap between electrodes and L is the electrode length (Figure 3.5 in Chapter I).

The $V_\pi$ equation also relates the frequency of RF signal and wavelength of the laser source to optical power of the EOM in Equation 1. Thus for study the frequency characterization of the EOM, this time the phase difference between RF signals at source and receiver is evaluated when the frequency of RF signal increase from 50 Hz to 2 MHz.

The frequency characterization of the CD setup is showing that under constant wavelength (1540 nm) of laser source, the phase difference between transmitter and receiver is quite sensitive to length of fiber.

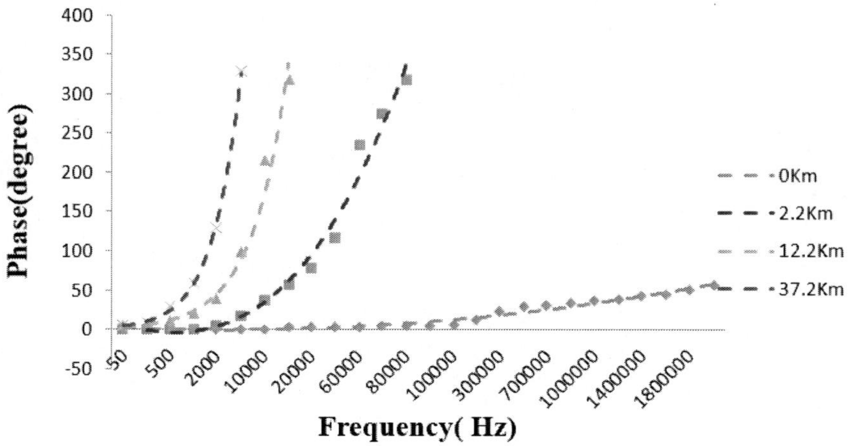


Figure 2.22. Characterization of CD measurement setup by measuring phase variation of RF signal at different RF frequency.

At distance of less than 1 km, the response of the setup to the frequency of the RF signal below than 1 MHz is almost insignificant, but Increasing of frequency to 2 GHz with in the short distance slight increase of the phase difference can be observed. At higher frequency and at longer distance the system becomes enormously sensitive to the small variation of RF signal. In fact, it has been expected that the system being insensitive to the variation of RF signal at any long haul distance, while at the longest fiber length the measurements reveals a remarkable variation of the phase at the frequency of less than 2 MHz. From Figure 22, it can conclude that the optimal frequency point for normal operation of EOM device and accurate dispersion measurement is 50 Hz or below than of it.

## 2. Analysis of the Dispersion Measurement Results

The results of CD measurement using phase difference method are obtained after complete characterization of EOM modulator. The final characterization of the setup is presented in Table 2.2. The dispersion and group delay measurements are evaluated for 10 km and 25 km of the fiber length. In computation of group delay and dispersion parameters, the Equation 2.3 and 2.4 has been used.

**Table 2.2. Optimal operation point for accurate CD measurement**

| | |
|---|---|
| EOM | 792000280 |
| Serial No | 7F6e1900 |
| $V_b$ | 8.5v |
| $V_p$ | 1.1v |
| Frequency | 50 Hz |

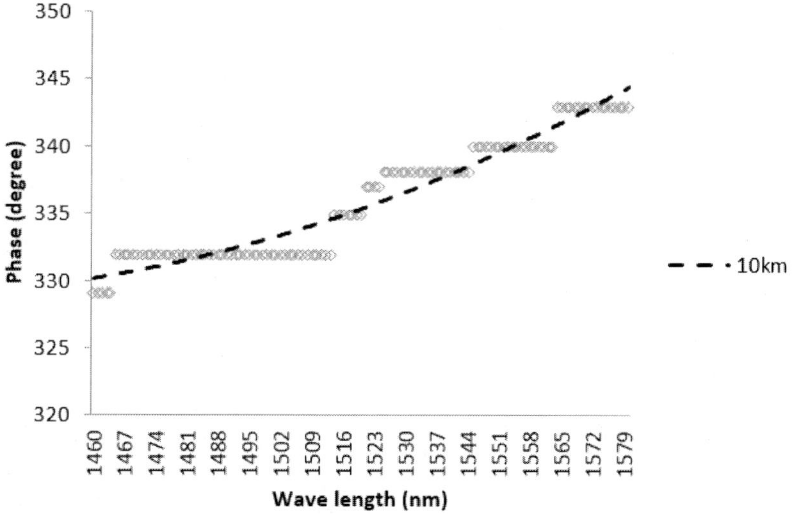


Figure 2.23. relative measured phase vs. wavelength for 10 km fibre.

The final results of CD measurements using phase shift technique is demonstrated in Figures 2.23-2.26. The shifts in measured phase obviously presents more difference correspond to fiber length of 25 Km rather than 10 Km (Figure 2.23 and 2.24). The remarkable increase of dispersion and group delay with increase of wavelength (1460nm-1580nm) can be observed after implementation of measured phase and wavelength values in Equations 2.3 and 2.4 and plotting them in respective Figures 2.25 and 2.26.

$$\Delta T_\lambda = -\frac{\varphi_{\left(\lambda+\frac{\Delta\lambda}{2}\right)} - \varphi_{\left(\lambda-\frac{\Delta\lambda}{2}\right)}}{360 f_m} \times 10^{12} \tag{2.3}$$

$$D_{\lambda_i} = \frac{\Delta\varphi_{\lambda_i} - \Delta\varphi'_{\lambda_i}}{360 f_m L \Delta\lambda} \times 10^{12} \tag{2.4}$$

where, $\lambda$ is the centre of wavelength interval, $f_m$is the modulation frequency in Hz and $\varphi$is the phase of recovered modulation signal and $\Delta\lambda$ wavelength interval between two adjacent wavelengths. The delay in terms of measured phase and wavelength is showing by $\Delta T_\lambda$ parameter while dispersion as function of wavelength presents by $D_{\lambda_i}$.

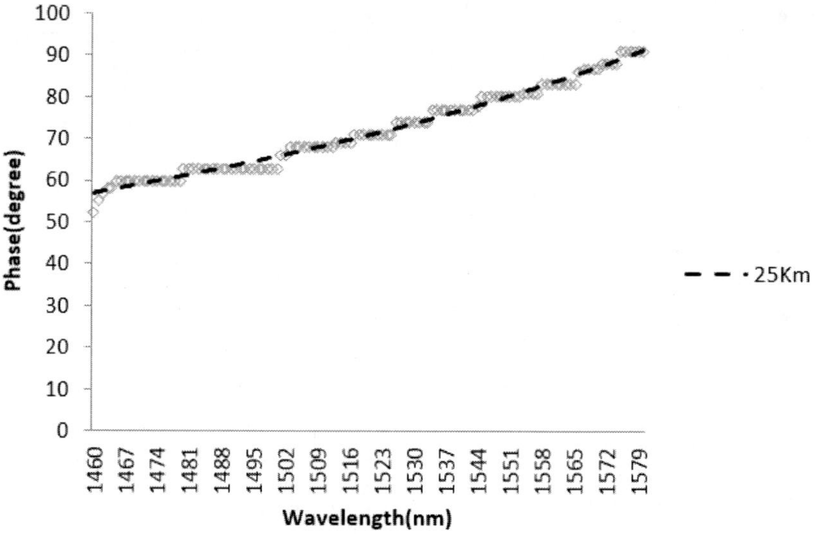


Figure 2.24. relative measured phase vs. wavelength for 25 km fibre.

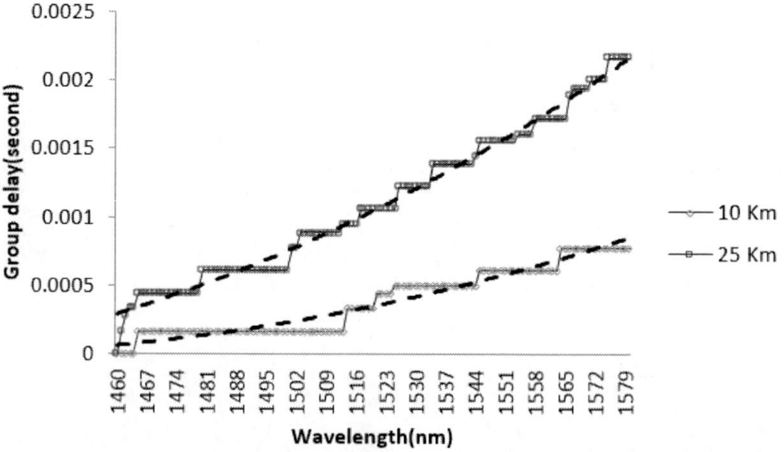


Figure 2.25. Group delay measurement vs. wavelength at different fibre length.

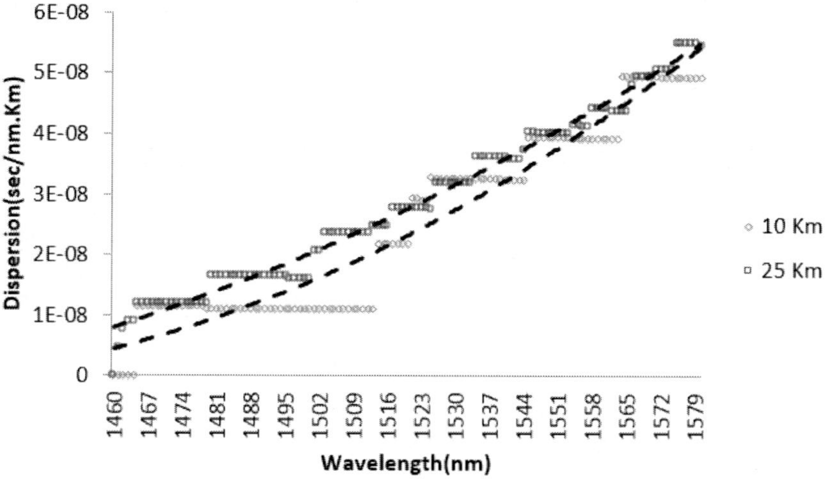


Figure 2.26. Dispersion measurement vs. wavelength at different fibre length.

# REFERENCES

[1] I. S. Amiri, S. Alavi, M. Soltanian, R. Penny, A. Supa'at, N. Fisal, and H. Ahmad, 2×2 MIMO-OFDM-RoF Generation and Transmission of Double V-Band Signals Using Microring Resonator System, *Optical and Quantum Electronics*, 2015.

[2] S. E. Alavi, I. S. Amiri, S. M. Idrus, A. S. M. Supa'at, J. Ali, and P. P. Yupapin, All Optical OFDM Generation for IEEE802.11a Based on Soliton Carriers Using MicroRing Resonators *IEEE Photonics Journal, 6(1)*, 2014.

[3] I. S. Amiri, S. E. Alavi, N. Fisal, A. S. M. Supa'at, and H. Ahmad, All-Optical Generation of Two IEEE802.11n Signals for 2×2 MIMO-RoF via MRR System, *IEEE Photonics Journal*, *6(6)*, 2014.

[4] I. S. Amiri and J. Ali, Data Signal Processing Via a Manchester Coding-Decoding Method Using Chaotic Signals Generated by a PANDA Ring Resonator, *Chinese Optics Letters*, *11(4)*, 2013, 041901(4).

[5] I. S. Amiri, A. Shahidinejad, A. Nikoukar, M. Ranjbar, J. Ali, and P. P. Yupapin, Digital Binary Codes Transmission via TDMA Networks Communication System Using Dark and Bright Optical Soliton, *GSTF Journal on Computing (joc)*, *2(1)*, 2012, 12.

[6] I. S. Amiri, S. E. Alavi, and J. Ali, High Capacity Soliton Transmission for Indoor and Outdoor Communications Using Integrated Ring

Resonators, *International Journal of Communication Systems*, *28(1)*, 2015, 147–160.

[7] I. S. Amiri, S. E. Alavi, Sevia M. Idrus, A. Nikoukar, and J. Ali, IEEE 802.15.3c WPAN Standard Using Millimeter Optical Soliton Pulse Generated By a Panda Ring Resonator, *IEEE Photonics Journal*, *5(5)*, 2013, 7901912.

[8] Iraj Sadegh Amiri, Ali Nikoukar, and Sayed Ehsan Alavi, Soliton and Radio over Fiber (RoF) Applications. Saarbrücken, Germany: LAP LAMBERT Academic Publishing, 2014.

[9] Iraj S. Amiri, Soliton-Based Microring Resonators: Generation and Application in Optical Communication. USA: Amazon, 2015.

[10] Sayed Ehsan Alavi, Iraj Sadegh Amiri, and Abu Sahmah M. Supa'at, Analysis of VFSO System Integrated With BPLC. Amazon: Lap Lambert Academic Publishing, 2014.

[11] I. S. Amiri, S. E. Alavi, and H. Ahmad, Analytical Treatment of the Ring Resonator Passive Systems and Bandwidth Characterization Using Directional Coupling Coefficients. *Journal of Computational and Theoretical Nanoscience (JCTN)*, *12(3)*, 2015, 418-424.

[12] I. Amiri, M. Soltanian, and H. Ahmad, Application of Microring Resonators (MRRs) in Optical Soliton Communications, in Optical Communication Systems: Fundamentals, Techniques and Applications, ed: Novascience Publisher, 2015, pp. 25 - 44.

[13] C. Tanaram, C. Teeka, R. Jomtarak, P. P. Yupapin, M. A. Jalil, I. S. Amiri, and J. Ali, ASK-to-PSK generation based on nonlinear microring resonators coupled to one MZI arm, *Procedia Engineering*, *8*, 2011, 432-435.

[14] I. S. Amiri, Attenuation Measurement of Transparent Materials and LIDAR Using NIR Laser. USA: Amazon, 2015.

[15] H Ahmad, M. R. K. Soltanian, I. S. Amiri, S. E. Alavi, A. R. Othman, and A. S. M. Supa'at, Carriers Generated by Mode-locked Laser to Increase Serviceable Channels in Radio over Free Space Optical Systems, *IEEE Photonics Journal*, *7(5)*, 2015.

[16] I. S. Amiri, S. Ghorbani, P. Naraei, and H. Ahmad, Chaotic Carrier Signal Generation and Quantum Transmission Along Fiber Optics Communication Using Integrated Ring Resonators, *Quantum Matter*, *4(2)*, 2015, 151-155.

In: Systems Engineering
Editor: Marlene Hopkins

ISBN: 978-1-63485-752-9


*Chapter 3*

# A REVIEW OF OPTICAL WAVEGUIDE SENSORS USING ION-EXCHANGE TECHNOLOGY

***I. S. Amiri, M. M. Ariannejad and H. Ahmad***
Photonics Research Centre, University of Malaya,
Kuala Lumpur, Malaysia

## ABSTRACT

Optical interconnects are expected to complement or even replace electrical interconnects for shorter and shorter distances, as they allow for large data transmission rates as well as very high link densities. An attractive way to implement optical interconnects in modern multiprocessor systems is to integrate an optical waveguide layer into conventional printed circuit boards. A technology based on ion-exchange in commercially available glass substrates developed, as a fabrication technology for optical waveguides in the context of optical interconnects. Glass has unprecedented properties that make it the material of choice for many optical applications. Silver ion-exchange in glass reveals to be a highly promising method for the fabrication of low loss optical waveguides, especially also in the context of optical interconnects, because it is a scalable, industrially compatible and low-cost technology.

**Keywords**: ion-exchange, optical sensors, propagation loss, refractive index, optical waveguide, glass

# 1. Introduction

The term 'integrated optics' created more than 30 years ago and describes a family of technologies where light guiding structures are integrated on planar substrates [1-12]. The concept corresponds to electronic integration, where lithographical tools are used to create structures in the micron range with submicron precision. One of the first reliable production processes for integrated optical waveguides was ion exchange in glass. The ion-exchange technique has been used for centuries to modify optical absorption properties of glasses for purposes of decoration and coloring [13]. However, only in the late 70's with the pioneering work of Izawa and Nakagome [14] and Gialorenzzi et al. [15]. It was showed that the technique is suitable to produce optical waveguides [16]. Since then, this fabrication process has been used in several applications of integrated optics: optical communications, signal processing and optical sensing [15, 17]. The advantages of the technique are low material and processing costs, low optical losses produced in the waveguide and compatibility with the optical fiber assembly. Typically, in this procedure, $Na^+$ ions present in the glass are exchanged with other ions from a melt salt. These ions have different ionic radius and polarizability from those of $Na^+$, inducing stress in the glass matrix, accompanied by modification of its refractive index [13]. The most common ion used is silver ($Ag^+$) due to its great variation in the refractive index compared to other ions, such as potassium ($K^+$) and thallium ($Tl^+$) [18]. In order to reduce the fabrication effort for integrated optic components, which become more and more complex at present, modelling procedures attain an increasing interest. The effective index method and beam propagation (BPM) calculations are commonly used tools for the prediction of waveguide properties and component parameters, respectively.

Glasses do not have fundamental material birefringence and they are amorphous, altered from semiconductors that is in crystalline form. Equally the shape of the waveguide and the stresses that rise during fabrication contribute to birefringence, nevertheless with suitable design, these can be well-adjusted together with each other to yield the single-mode devices that has little polarization dependence [19]. Furthermore, ion exchange has other methods that exist concluded to glass waveguides fabricated. Most include the deposition of thin glass films (e.g., sol-gel coating, flame hydrolysis deposition, chemical vapor deposition), by using reactive ion etching to define the device geometry, and subsequent deposition of the over cladding. Several deposition steps and etching can be variety of these procedures time consuming costly. Ion exchanged waveguides possess may have the required characteristics which is stated in Table 1.

The highest inspirations of the usage in ion exchange with glass was optical waveguide sensors, which are the resulting to: glass has insignificant is rough against a diversity and optical attenuation, thermal strains and mechanical and.

**Table 1. Comparison of silica-on-silicon, polymer and ion-exchange (from salt melts) Waveguide technologies for optical waveguide fabrication**

| Properties | Silica-on-Silicon | Polymer | Ion-exchange |
|---|---|---|---|
| Coupling to optical fiber | Fair | poor | excellent |
| Elimination of lenses | poor | Poor | Good |
| Propagation loss | excellent | Fair | Good |
| Polarization sensitivity | Fair | Fair | excellent |
| Thermal stability | Good | poor | Good |
| Tapering (coupling to laser) | Fair | - | excellent |
| Miniaturization of optical devices | excellent | Good | Fair |
| Cost saving - materials | Fair | Good | excellent |
| Cost saving - process | poor | Fair | excellent |
| Production yield | fair/poor | - | excellent |
| Number of fabrication steps | poor | Poor/fair | excellent |
| Packaging ease | fair/poor | fair | excellent |

## 2. Experimental

The ion-exchange technique is a very old technique that used by the Egyptians in the 6$^{th}$ century to decorate dishes and pots. In the Middle Ages and later during the Renaissance, glass staining was a widely used method to fabricate colored church windows. Probably the best known are the yellow ("Jaune d'Argent") and amber silver stains, the red copper (copper-ruby) and red gold (gold-ruby) or purple gold ("Purpleof Cassius") stains. First industrial applications, however, emerged only in the beginning of the 20$^{th}$ century, with chemical surface tempering of glass (also referred to as ion-stuffing or glass-strengthening), i.e., a large ion from an alkali salt melt (e.g., $K^+$) is exchanged with a smaller ion in the glass (e.g., $Na^+$), thus creating a compressive stress state at the glass surface. Initial studies of ion-exchange processes and ion conduction processes in glass are presented. In 1918, Schott researchers observed that ion-exchange also produces a change in the refractive index of such a diffused layer. After the beginning of the integrated optics era, in the late 1960s, research was focused on the development of thin-film optical components like power splitter, couplers, modulators and soon, interconnected on a single chip. Pioneering work was done, in the early 1970s, by Izawa et al. [20]. They opened the way to extended activities in the research and development of the ion-exchange technique in glass [16]. A comprehensive range of optical waveguide properties can be done by using ion-exchange method. Equally, weakly and strongly guiding single mode waveguides, but also large multimode waveguides can be expected. Surface guides in addition to buried waveguides have been processed. In the last years, considerable effort has been put into the fabrication of waveguides in rare-earth (e.g., $Er^{3+}$, $Yb^{3+}$) doped silicate as well as non-silicate glass substrates to achieve optical amplification and hence active optical devices [21-23].

## 3. THERMAL ION-EXCHANGE

In a certain temperatures and concentration gradient of similar ions at the glass surface could generate the movement of ions. The exchange of mobile charged kind happens on a one basis to reserve the glass to be neutral. The amount of ion exchange controlled with slower mobile ions [24].

The glass substrate that modify ion like $A^+$ that could be dipped in salt that is molten with the present of the cation $B^+$ like $Ag^+$, $K^+$, $Tl^+$ as shown in Figure 1. The interface of the glass melt, both ions shortly could decrease to zero. The thermal will cause the interface for $B^+$ ions to exchange with the $A^+$ ions and similarly opposite. Moreover, the cations that diffuse can be away from interface. The freedom of movement, $A^+$ cations in the molten salt existence is higher, they move fast vanished from the interface, while the $B^+$ cations gradually exchange into the glass, generating a layer near the glass surface that is thin. At glass surface, the concentration of exchanged $B^+$ cations is reductions monotonically and highest inside the glass. The procedure could be fast within the high temperatures, since the rise of ion mobility through considering the temperature, however the glass structure relaxes with the rise of temperature.

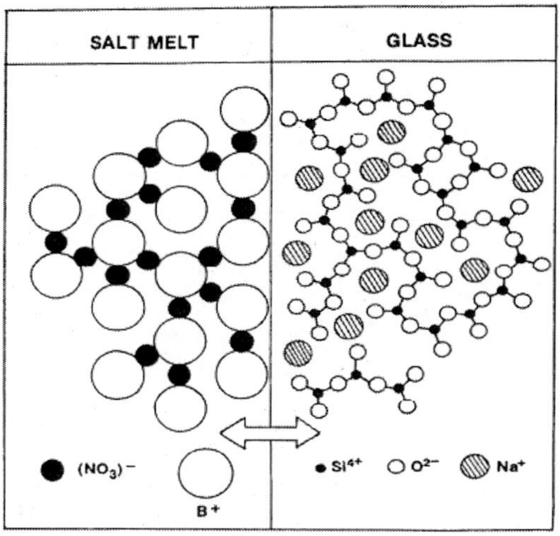


Figure 1. Illustration of the melt-glass interface (here $A^+ = Na^+$). $B^+$ is the cation of the melt (e.g., $Ag^+$, $K^+$, $Tl^+$,...) which will be exchanged with the cation in the glass. The melts usually consist of nitrates, chlorides or sulphates.

Table 3 the suitable factors in case of generally used ions for ion-exchange procedures: ionic radius, electronic polarizability, corresponding melting and normally with the usage of salts or the induced refractive index change and also decomposition point, birefringence and propagation loss after ion-exchange [25].

Silver ion-exchange is an appropriate method for $Ag^+$-ions from the $AgNO_3$ melt exchanged with mixed ions $Na^+$ in the glass, though not high temperature (usually between 220 till 300°C). $\Delta n$ is the refractive index that can change around 0.1, soda-lime glasses that can achieve to 0.22 for TiF6 phosphate-glass (Schott Glaswerke, Mainz) with the usage of thermal in ion-exchange with 410°C temperature [26].

Refractive index would Decreasing by increasing the usage of diluted salt melts. Najafi et al. [27] discussed a method in that concluded electrolytic issued of silver ions from a silver rod, particular control in case of silver concentration in the $Ag^+$ and $Na^+$ which is melt above the method that described, consequently with considering the waveguides fabrication and reproducibility (the reproducibility can be approximately ~ $10^{-4}$ in case of continuous propagation). The significant consideration in silver exchange as point out before, is the decrease of silver, that happened at higher temperatures and is maintained by decreasing the ions like $Fe^{2+}$, $As^{3+}$, S $b^{3+}$ and (acting as electron donors) present in the glass as the layers [28].

**Table 3. Parameters for ion-exchange in glass**

| Ion | Ionic radius [$\dot{A}$] | Electronic polarizability α [$\dot{A}^3$] [104] ([105]) | Salt | Melting (decomp.) point [°C] | Refractive index change Δn | Birefringence $\delta n \times 10^4$ | Propagation loss [dB/cm] | Comments |
|---|---|---|---|---|---|---|---|---|
| $Li^+$ | 0.65 | 0.03 (0.029) | $LiNO_3$ | 264 (600) | 0.012 | 2 | >1 | Tensile stresses and surface crystallization |
| $Na^+$ | 0.95 | 0.41 (0.24) | $NaNO_3$ | 307 (380) | -0.02-0.002 | - | - | - |
| $K^+$ | 1.33 | 1.33 (0.9) | $KNO_3$ | 334 (400) | 0.009 | 2-20 | <0.2 | Compressive stresses |
| $Rb^+$ | 1.49 | 1.98 (1.4) | $RbNO_3$ | 310 (-) | 0.015-0.02 | 2-10 | >1 | Expensive |
| $Cs^+$ | 1.65 | 3.34 (2.44) | $CsNO_3$ | 414 (-) | 0.04 | 2 | 0.1 | Slow diffusion |
| $Ag^+$ | 1.26 | 2.4 (1.89) | $AgNO_3$ | 212 (444) | 0.1- 0.22 | 0-10 | <0.2 | Low thermal stability and photo-stability |
| $Ti^+$ | 1.49 | 5.2 (3.88) | $TiNO_3$ | 206 (430) | 0.1-0.2 | 0-50 | <0.1 | High toxcity |
| $Cu^+$ | 0.96 | 1.73 (1.22) | CuCl | 430 (-) | 0.015-0.06 | - | - | Risk of coloration ($Cu^{2+}$) |

**Table 4. Several loss data for silver ion-exchanged or potassium and silver double exchanged waveguides (from melts) partly post-treated**

| Fabrication method and/or post-treatent | Propagation loss [dB/cm] | Wavelength [nm] | Glass type | Reference |
|---|---|---|---|---|
| Ag-Na exchange K-Na/ Ag-Na double exchange (ion-masking) | 1.3 ~0.25 | 1296 | Corning 0211 | [32] |
| K-Na/Ag-Na double exchange and 24 h annealing | 0.3 | 1550 | Schott Bk7 | [33] |
| Ag-Na (buried waveguide) | ~0.15 | 1550 | Schott BGG 31 | [34] |

The losses depend significantly on the wavelength that is used and have a maximum above 400 nm that resonance for the precipitated small silver particles happens. The loss surge regarding to colloidal of silver much smaller at wavelengths >800 nm [29]. The absorption losses are different from less than 0.1 dB/cm for BK7 glass (waveguides annealed at 500°C in air after the ion-exchange [30]) to a few dB/cm for soda-lime glasses in addition to Pyrex glass (the borosilicate glass not improved with admiration to impurities). Table 4 shows a few loss data of silver ion-exchanged waveguides. It is very tough to compare the altered available loss values, since it has altered processing conditions, ion-exchange, glass substrates, material systems and methods used, the various waveguide dimensions produced and also the optical characterization methods that are engaged. Potassium and silver double exchanged waveguides, which declared as an ionic masking procedure to have improved loss figures at the outflow of a more complex procedure (improved number of process-steps). Furthermore, the annealing in addition to the burying waveguides characteristically declines the propagation losses considerably.

The structure of the sensor, that the operation can be defined in terms of coupled modes, is illustrated in Figure 1. A single TM mode is excited in the input waveguide and divided into two paths with a 3 dB y-junction splitter. The mode propagating under the gold film then couples to the surface plasmon mode, guided by the interface between the metal film and the superstrate, if the two modes are closely phase matched. If a thin film is adsorbed to the metal surface then the coupling condition between the two modes is altered causing a change in the output intensity of the sensor. The other arm of the y-junction acts as a reference to minimize the effects of input power drifts.

Extensive design work [31] has led to optimized sensor designs for the aqueous environment.

Lesser process temperature could reduce the silver to decrease the properties, eutectic melts that can be used [35, 36] in burying Waveguides, which significantly decreases optical losses (creating primarily from surface scattering) in addition to polarization requirement of the waveguides, to obtain by a second ion-exchange in a $NaNO_3$ melt along with usage of eutectic melt of $NaNO_3$ and $KNO_3$. Reduce the melting point of the eutectic melts can cause the profile rise of the waveguides because the decreasing of thermal diffusion. Lastly, it has also to be declared that the refractive index at the glass surface induced by silver ion-exchange from melts shows a dependence on the melt concentration, also varies for altered glasses (e.g., [37] and Table 2). Also, a robust procedure temperature as well as period requirement was observed in [38]. Currently Capek et al. [39] and Salavcova et al. [85] evaluated the ion-exchange ($Ag^+$ and $K^+$) from a glass compositional point of view by altering the absorption of the intermediary oxides (CaO, MgO and ZnO). The ion-exchange depth is improved when the transitional oxide advances from ionic to more covalent bonding; that is from compositions containing CaO and MgO to compositions containing only ZnO as the intermediate oxide. Therefore, the more covalent character of the glass sorts its structure additional open for the powerful particles (i.e., the exchanging ions). The ionic character of the modifiers ($Ca^{2+}$ and $Mg^{2+}$) means quite strong electrostatic services among the non-bridging oxygen (NBO) and the neighboring $Na^+$ that impede the measure of the sodium ion through the glass matrix, therefore slowing down the exchange of the $Na^+$ for additional ion. Consequently, for the choice of a tolerable substrate for ion-exchange, a glass will be chosen, which preferably is free of MgO and CaO. Correspondingly, the number of non-bridging oxygens (NBO) shall be as low as possible. Ion-exchange from silver has been studied by numerous researchers [40-42]. To summarize, the benefits of such a solid-state procedure within the molten salt and using in ion-exchange are,

- Pure materials deposited by sputtering, evaporation,
- The variety of exchange temperatures like assisted field process that permits ion exchange at not high temperatures
- Ion-exchange would happened with electric field, not ion-exchange complete with heating up (minor thermal dispersion from the source)
- Constant ion concentration profile relevant to ion source
- Dry procedure can be appropriate in case of mass manufacture and not exactly electrically insulated and sample holders essential permitting

with the usage of high temperature in the process (in the occasion for ion-exchange from melts)

- Charge controlled procedure - existing monitoring permits for intrinsic procedure control
- Within a lot variety change in temperatures like field assisted procedure it permits to ion exchange to be happened in low temperatures (inadequate by melting and/or decomposition temperatures of salts)

To determine refractive index alteration Δn was comparable to molten salt in ion exchange since alike ions exchanged. Though, requirement of the surface refractive index on the functional electric field [43]. It can be promising to fabricate the waveguides in multimode and using relatively small fields and short distribution times. In multimode silver-film ion-exchanged waveguides, losses below 0.1 dB/cm for slab and 0.2 dB/cm for channel waveguides have been described [44]. Multimode waveguides (10 guided modes) have been established by Najafi et al. [43] in soda-lime glass (Fisher Premium with 14.31 wt% of $Na_2O$), with T = 275°C, E = 15 V/mm and t = 30 min. These multimode guides presented 0.5 dB/cm although singlemode waveguides presented 1 dB/cm absorption losses. Honkanen et al. [140] established multimode waveguides in Corning 0211 borosilicate glass for numerous ion streams, temperatures and development times. In case of fabrication in multimode waveguides it can drive the ions rapidly into the glass (this first step defines the refractive index alteration and the depth of the waveguide as shown in Figure 2) and achieve a following thermal or field-assisted step to variation the shape of the refractive index profile. All ion exchange processes are described here as they pertain to the silver-sodium ($Ag^+$–$Na^+$) system of ions. Figure 2 illustrates the thermal exchange from a molten salt source through a metal-oxide mask. The mask is used to define the device geometry for the exchanges from molten salt. $Ag^+$ ions are driven into the substrate from the melt by a chemical potential gradient, and in order to preserve charge neutrality, $Na^+$ ions are released into the melt [45]. Once in the glass phase, $Ag^+$ ions are-redistributed by thermal diffusion. The dielectric mask serves to reduce the occurrence of metallic silver clusters which would otherwise form due to electrolytic deposition at the glass/mask interface [40]. Such clusters produce a spectral attenuation in the waveguides. Thermal exchange may also be assisted by an applied electric field.

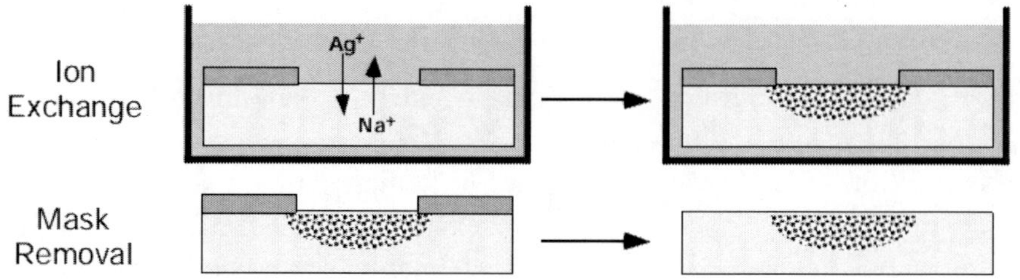


Figure 2. The four most common processing configurations for ion exchange. (a) thermal exchange from a molten salt.

## 4. Fabrication of the Glass Chip Sensor

The applications of guided-wave optical biosensors, and their market possibilities, are well 'documented.' Many optical transducer mechanisms are possible and they are currently generating a great deal of excitement as the potential of such devices is realized in the laboratory and, increasingly, commercially. Optical biosensors are generally small, light, and rugged, offering the portability required for a field monitoring system. Integrated optical sensors maybe connected by optical fibers and allow for the fabrication of multiple sensors in a single substrate using photolithographic techniques. The exploitation of a surface Plasmon resonance (SPR) to form the basic transduction mechanism of a chemical sensor has been known for some years. A surface Plasmon is usually a transverse magnetic (TM) electromagnetic mode guided by the interface between two media whose dielectric constants have real parts of opposite sign. For visible light this requirement can be fulfilled by a dielectric and a metal. The formation of a surface Plasmon can be achieved by using a 'bulk' optical component such as a prism, and equipment employing this approach is now commercially available. Another option is to employ the distributed coupling between a planar waveguide and a surface Plasmon in a metal-coated waveguide. The use of SPR transduction in an optical sensor allows for the use of the metal film as an electrode for electrochemical control of sensing reactions. The general configuration of the sensor, whose operation can be described in terms of coupled modes, is detailed in Figure 3. A single TM mode is excited in the input waveguide and divided into two paths with a 3 dB Y-junction splitter. The mode propagating under the gold film then couples to the surface Plasmon mode, guided by the interface between the metal layer and the dielectric superstrate, if the two modes are closely phase matched. If a thin dielectric film is adsorbed to the metal surface then the effective index of the surface Plasmon mode will be altered, changing the coupling condition between the waveguide and surface Plasmon modes. This change can then be monitored, for example, as a function of wavelength or a change in the output intensity of the sensor. The other output of the Y-junction splitter is used as a reference channel to minimize the effects of input power drifts [46]. Figure 3 details the process flow of the glass sensor fabrication. After cleaning the glass wafer, a 100 nm film of alumina is deposited by sputtering [47]. The waveguide pattern is then transferred on the mask by photolithography, followed by a wet etching mask opening of the alumina film.

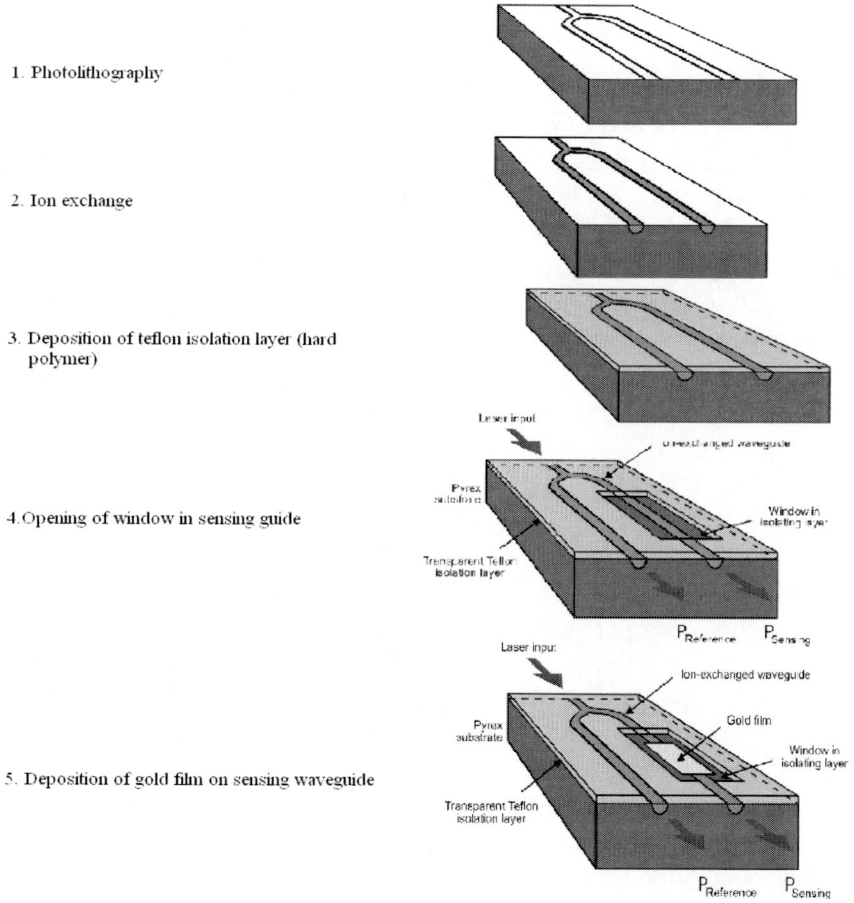


Figure 3. Process flow of the glass sensor fabrication.

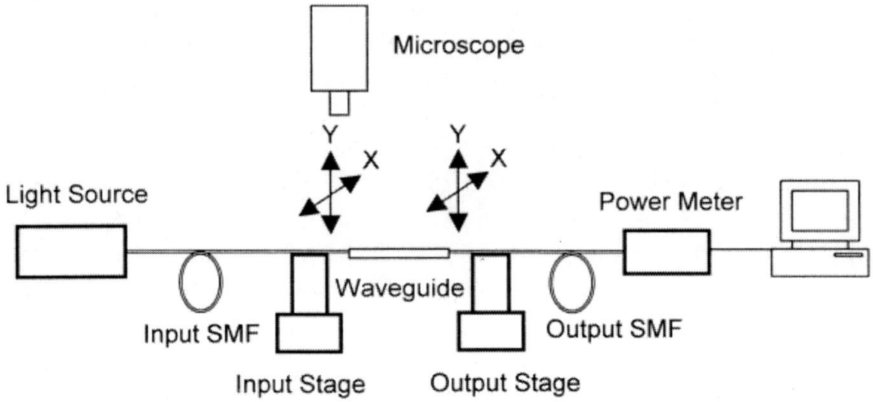


Figure 4. Waveguide-characterization setup.

The characterization of the waveguides was carried out with the Newport Auto Align Packaging system shown in Figure 4. The light source is a broadband laser diode module with a wavelength range of 1.53-1.61 μm. Both the input and the output fibers were attached to a computer-controlled stage with linear positioning accuracy of 0.05 μm. The input and out-put stages can be controlled to scan along both the lateral X and the vertical Y directions to find the optimal coupling positions between the fiber and the waveguide [48]. A microscope at the top was used as a monitor to keep a minimum gap between the end-faces of the fiber and the waveguide.

## Conclusion

Ion-exchange is one of the techniques for over 30 years nowadays and characterizes a matured and it is purposed to use in optical waveguide fabrication method, mainly for scale down the waveguides, integrated optical devices that is compacted. The enactments of such optical devices even exceed the Telcordia (GR 1221) that it can be used in telecommunication [21].

Altered ion-exchange procedures reviewed, the ion-exchange that is film can be appeared to be a good solution for sensing uses. This is primarily regarding to significant explanation of the processing setup (electrically insulated sample holders and no high-temperature leak-proof needed) in addition to the characteristic control of the origination of the ion-exchange (no ions produced without relating an electric field) and ion-migration concluded current monitoring. The silver ion-exchange offerings the properties of an scientifically well-matched and scalable procedure, appropriate for mass manufacture. Gold-coated waveguide SPR sensors is fabricated and used to monitor the attachment of biotin-avidin layers to the surface of the sensors in the aqueous environment. Subsequently the gold films of the sensors were chemically modified, and when used in conjunction with appropriate antibodies, enable the detection of the pesticide simazine. Utilizing the gold film as an electrode to allow the possibility of performing electrochemical experiments also remains to be investigated and is a potential advantage over other, dielectric based, integrated optical biosensors.

## REFERENCES


[1] SE Alavi, IS Amiri, H Ahmad, ASM Supa'at, and N Fisal, Generation and Transmission of 3 × 3 W-Band MIMO-OFDM-RoF Signals Using Micro-Ring Resonators, *Applied Optics*, 53(34), 2014, 8049-8054.

[2] IS Amiri, SE Alavi, and H Ahmad, Fiber laser setup used to generate several Mode-Locked pulses applied to soliton-based optical transmission link, In: *Horizons in World Physics*. vol. 287, ed. US: Novascience, 2015.

[3] IS Amiri, SE Alavi, and H Ahmad, Optically generation and transmission ultra-wideband mode-locked lasers using dual-wavelength fiber laser and microring resonator system, In: *Horizons in World Physics*. vol. 287, ed. US: Novascience, 2015.

[4] IS Amiri, SE Alavi, and H. Ahmad, Microring resonators used to gain the capacity in a high performance hybrid wavelength division multiplexing system, In: *Horizons in World Physics*. vol. 287, ed. US: Novascience, 2015.

[5] H Ahmad, MRK Soltanian, IS Amiri, SE Alavi, AR Othman, and ASM Supa'at, Carriers Generated by Mode-locked Laser to Increase Serviceable Channels in Radio over Free Space Optical Systems, *IEEE Photonics Journal*, 7(5), 2015.

[6] IS Amiri, S. E. Alavi, Sevia M. Idrus, A. Nikoukar, and J. Ali, IEEE 802.15.3c WPAN Standard Using Millimeter Optical Soliton Pulse Generated By a Panda Ring Resonator, *IEEE Photonics Journal*, 5(5), 2013, 7901912.

[7] MRK Soltanian, IS Amiri, WY Chong, SE Alavi, and H Ahmad, Stable dual-wavelength coherent source with tunable wavelength spacing

generated by spectral slicing a mode-locked laser using microring resonator, *IEEE Photonics Journal*, 7(6), 2015.

[8] S. E. Alavi, I. S. Amiri, S. M. Idrus, ASM Supa'at, J. Ali, and P. P. Yupapin, All Optical OFDM Generation for IEEE802.11a Based on Soliton Carriers Using MicroRing Resonators, *IEEE Photonics Journal*, 6(1), 2014.

[9] IS Amiri, S. Alavi, M. Soltanian, R. Penny, A. Supa'at, N. Fisal, and H Ahmad, 2 × 2 MIMO-OFDM-RoF Generation and Transmission of Double V-Band Signals Using Microring Resonator System, *Optical and Quantum Electronics*, 2015.

[10] IS Amiri, SE Alavi, MRK Soltanian, N Fisal, ASM Supa'at, and H Ahmad, Increment of Access Points in Integrated System of Wavelength Division Multiplexed Passive Optical Network Radio over Fiber, *Scientific Reports*, 5, 2015.

[11] MRK Soltanian, H Ahmad, A Khodaie, IS Amiri, MF Ismail, and SW Harun, A Stable Dual-wavelength Thulium-doped Fiber Laser at 1.9 μm Using Photonic Crystal Fiber, *Scientific Reports*, 2015.

[12] SE Alavi, MRK Soltanian, IS Amiri, M Khalily, ASM Supa'at, and H Ahmad, Towards 5G: A Photonic Based Millimeter Wave Signal Generation for Applying in 5G Access Fronthaul, *Scientific Reports*, 6, 2016.

[13] G. C. Righini and S. Pelli, Ion exchange in glass: a mature technology for photonic devices, In: *International Symposium on Optical Science and Technology*, 2001, 93-99.

[14] Y. Hao, H. Wang, B. Zheng, Y. Li, J. Yang, X. Jiang, Q. Zhou, and M. Wang, Two layers of buried optical waveguides in glass substrate by ion exchange, *Optical Engineering*, 51(4), 2012, 044603-1-044603-4.

[15] A. Cosentino, Q. Tan, M. Roussey, and H. P. Herzig, Planar Integrated Sensors on Waveguides for sensing applications, 2011.

[16] P. L. Inácio, B. J. Barreto, F. Horowitz, R. R. Correia, and M. B. Pereira, Silver migration at the surface of ion-exchange waveguides: a plasmonic template, *Optical Materials Express*, 3(3), 2013, 390-399.

[17] V. Donzella, V. Toccafondo, S. Faralli, F. Di Pasquale, C. Cassagnettes, D. Barbier, and H. H. Figueroa, Ion-exchanged Er 3+/Yb 3+ co-doped waveguide amplifiers longitudinally pumped by broad area lasers, *Optics Express*, 18(12), 2010, 12690-12701.

[18] K. Kimura, S. Sakida, Y. Benino, and T. Nanba, Fabrication and characterization of Er3+-doped tellurite glass waveguide by Ag+-Na+

ion-exchange method, In: *IOP Conference Series: Materials Science and Engineering*, 2011, 112018.

[19] P. Chen, Y. Zhu, Y. Shi, D. Dai, and S. He, Fabrication and characterization of suspended SiO2 ridge optical waveguides and the devices, *Optics Express*, 20(20), 2012, 22531-22536.

[20] S. Karlsson, "Modification of Float Glass Surfaces by Ion Exchange," LNU, 2012.

[21] I. Artundo, A. Tymecki, E. Ortego, and B. Ortega, Cost forecasting of passive components for optical fiber network deployments, *Optical Fiber Technology*, 17(3), 2011, 218-226.

[22] H. Schroder, L. Brusberg, N. Arndt-Staufenbiel, J. Hofmann, and S. Marx, Glass panel processing for electrical and optical packaging, in *Electronic Components and Technology Conference (ECTC), 2011 IEEE 61st*, 2011, 625-633.

[23] O. Peña-Rodríguez, J. Olivares, M. Carrascosa, Á. García-Cabañes, A. Rivera, and F. Agulló-López, Optical Waveguides Fabricated by Ion Implantation/Irradiation: A Review, *Ion Implantation, Prof. Mark Goorsky (Ed.), ISBN*, 2012, 978-953.

[24] S. Wackerow and A. Abdolvand, Optical analyses of the formation of a silver nanoparticle-containing layer in glass, *Optics Express*, 20(21), 2012, 23227-23234.

[25] A. Tervonen, B. R. West, and S. Honkanen, Ion-exchanged glass waveguide technology: a review, *Optical Engineering*, 50(7), 2011, 071 107-071107-15.

[26] T. Kuhler and E. Griese, Modeling the ion-exchange process to support the manufacturing of optical multimode graded-index waveguides in thin glass sheets, In: *Signal Propagation on Interconnects (SPI), 2010 IEEE 14th Workshop on*, 2010, 87-89.

[27] V. A. G. Rivera, Ion Exchange to Fabrication of Waveguides for Optical Telecommunication, In: *Ion Exchange Technology I*, ed: Springer, 2012, pp. 467-490.

[28] S. Wackerow, G. Seifert, and A. Abdolvand, Homogenous silver-doped nanocomposite glass, *Optical Materials Express*, 1(7), 2011, 1224-1231.

[29] X. Han, F. Chu, F. Pang, H. Cai, R. Qu, and Z. Fang, Protection of phosphate glass by BK7 glass film in ion-exchange processing, *Optik - International Journal for Light and Electron Optics*, 120(8), 2009, 395-398.

[30] Y. Xiao, J. Yang, and M. Wang, Manufacturing of Glass-Based Optical Power Splitter by Ion-exchange Method, In: *Proc. of SPIE* Vol, 2012, 84162I-1.

[31] J. Homola, H. B. Lu, G. G. Nenninger, J. Dostálek, and S. S. Yee, A novel multichannel surface plasmon resonance biosensor, *Sensors and Actuators B: Chemical*, 76(1), 2001, 403-410.

[32] W. Jin, K. Chiang, and Q. Liu, Electro-optic long-period waveguide gratings in lithium niobate, *Optics Express*, 16(25), 2008, 20409-20417.

[33] J. Zou and R. T. Chen, Improvement of two-step K+–Na+ and Ag+–Na+ ion-exchanged glass waveguides by surface removal, *Applied physics letters*, 88(8), 2006, 081102-081102-3.

[34] S. Yliniemi, B. R. West, and S. Honkanen, Ion-exchanged glass waveguides with low birefringence for a broad range of waveguide widths, *Applied optics*, 44(16), 2005, 3358-3363.

[35] Y. Cheng, K. Sugioka, K. Midorikawa, and Z. Xu, Microbiochips Monolithically Integrated with Microfluidics, Micromechanics, Photonics, and Electronics by 3D Femtosecond Laser Direct Writing, *Microfluidics: Theory And Applications*, 2010, 1.

[36] L. Sansoni, F. Sciarrino, G. Vallone, P. Mataloni, A. Crespi, R. Ramponi, and R. Osellame, Polarization entangled state measurement on a chip, *Physical review letters*, 105(20), 2010, 200503.

[37] V. Nazabal, M. Poulain, M. Olivier, P. Pirasteh, P. Camy, J. L. Doualan, S. Guy, T. Djouama, A. Boutarfaia, and J. L. Adam, Fluoride and oxyfluoride glasses for optical applications, *Journal of Fluorine Chemistry*, 134(0), 2012, 18-23.

[38] R. Rogoziński, Ion Exchange in Glass–The Changes of Glass Refraction, 2012.

[39] G. Righini, S. Pelli, M. Brenci, M. Ferrari, C. Duverger, M. Montagna, and R. Dall'Igna, Active optical waveguides based on Er-and Er/Yb-doped silicate glasses, *Journal of non-crystalline solids*, 284(1), 2001, 223-229.

[40] I.E. Araci, R. Himmelhuber, C. T. DeRose, J. Luo, A. Y. Jen, R. Norwood, and N. Peyghambarian, Alignment-free fabrication of a hybrid electro-optic polymer/ion-exchange glass coplanar modulator, *Optics Express*, 18(20), 2010, 21038-21046.

[41] E. Cattaruzza, G. Battaglin, F. Gonella, A. Quaranta, G. Mariotto, C. Sada, and S. Ali, Chromium doping of silicate glasses by field-assisted solid-state ion exchange, *Journal of non-crystalline solids*, 357(8-9), 2011, 1846-1850.

[42] I.E. Araci, *Low-loss, high performance hybrid photonics devices enabled by ion-exchanged glass waveguides*, 2010.

[43] S. I. Najafi, Introduction to glass integrated optics: Artech House Norwood, 1992.

[44] Y. Chen, J. Jaakola, Y. Ge, A. Säynätjoki, A. Tervonen, S.-P. Hannula, and S. Honkanen, In situ fabrication of waveguide-compatible glass-embedded silver nanoparticle patterns by masked ion-exchange process, *Journal of non-crystalline solids*, 355(45), 2009, 2224-2227.

[45] S. Honkanen, B. R. West, S. Yliniemi, P. Madasamy, M. Morrell, J. Auxier, A. Schulzgen, N. Peyghambarian, J. Carriere, and J. Frantz, Recent advances in ion exchanged glass waveguides and devices, *Physics and Chemistry of Glasses-European Journal of Glass Science and Technology Part B*, 47(2), 2006, 110-120.

[46] R. Levy and S. Ruschin, SPR waveguide sensor based on transition of modes at abrupt discontinuity, *Sensors and Actuators B: Chemical*, 124 (2), 2007, 459-465.

[47] Schimpf, D. Bucci, M. Nannini, A. Magnaldo, L. Couston, and J.-E. Broquin, Photothermal microfluidic sensor based on an integrated Young interferometer made by ion exchange in glass, *Sensors and Actuators B: Chemical*, 163(1), 2012, 29-37.

[48] J. Zou, F. Zhao, and R. T. Chen, Two-step K+-Na+ and Ag+-Na+ ion-exchanged glass waveguides for C-band applications, *Applied optics*, 41 (36), 2002, 7620-7626.

In: Systems Engineering
Editor: Marlene Hopkins
ISBN: 978-1-63485-752-9


*Chapter 4*

# SILVER NANO PARTICLE ION-EXCHANGED GLASS WAVEGUIDE TECHNOLOGY

***I. S. Amiri, M. M. Ariannejad and H. Ahmad***
Photonics Research Centre, University of Malaya,
Kuala Lumpur, Malaysia

## ABSTRACT

Ion exchange in glass and key developments in the first years of research are briefly described. Ion exchange in glass is a well-established method for fabrication of passive and active integrated photonic devices. An overview of fabrication, characterization and modeling of waveguides is given and the most important waveguide devices and their applications are discussed. Ion exchanged waveguide technology has served as an available platform for studies of general waveguide properties, integrated optics structures and devices, as well as applications. It is also a commercial fabrication technology for both passive and active waveguide components.

The generation of nano sized metal particles in amorphous matrix, the changes of the optical properties and the intensity of the reflective index were analyzed based on Surface Plasmon Resonance (SPR). Surface plasmons, particularly localized surface plasmons in the case of metal nanoparticles, explain the properties from the viewpoint of physical phenomenon and present the changes of the electromagnetic field due to the presence of metal particles.

**Keywords:** ion-exchange, Nano particle, optical sensors, propagation loss, refractive index, optical waveguide, glass

# 1. INTRODUCTION

In recent years, plasmonics as a subfield of nano photonics has appeared and developed rapidly. It primarily focuses on the plasmon coupling effects within close proximity along metal nanostructures and on the manipulation of light at the nano scale based on the properties of propagating and localized surface plasmons [1, 2]. Two promising applications for plasmonics include [3]:

- plasmonic waveguides to guide electromagnetic energy below the diffraction limit along for example continuous metal strips or the chains of closely spaced metal nanoparticles
- plasmonic sensors to realize various types of sensing schemes based on surface plasmon resonances, such as refractive index plasmon sensing [4], surface enhanced Raman spectroscopy (SERS), and metal enhanced fluorescence [5].

Plasmonic biosensing deals with a broad range of applicability of using surface plasmon waves. The simplest and most popular biosensing application is refractive-index plasmon sensing, which has been widely used to monitor the analyte-surface binding interactions. It utilizes the evanescent field of a special mode of an electromagnetic field-the surface plasmon to measure the change of refractive index that occurs in response to the analyte binding at or near a metal surface [6].

Propagating and localized surface plasmon resonances (SPR) are both utilized in this application the propagating SPR at flat metal/dielectric boundaries with metal strips and nanowires, and the localized SPR around metal nanoparticle structures [7-10]. Another important application, which is utilized as the characterization method in this work, is the surface enhanced Raman scattering (SERS) [10-12]. Both periodic and random metal nanostructures are fabricated to enhance local electromagnetic fields near the surface of the structures and to obtain an intense Raman signal. In addition, the mechanism in surface enhanced fluorescence is analogous to that in SERS, by utilizing the enhanced local electromagnetic fields near the surface of the nanoparticle to increase the final quantum yield [13].

## 2. Molten Salt Ion Exchange

Glass is an interesting substrate material for integrated optics because of its relatively low cost, excellent transparency, and high threshold to optical damage. Considering the sensor applications, glass is also a relatively inert material to the bio-molecules. Ion exchange is a well-known technique proposed and developed since 1970s, to modify the electrical and optical properties of glass and to fabricate integrated optical devices such as waveguides. In the ion exchange process, some of the ions in the glass matrix (usually Na+) are replaced by the ions of the same valence from the surrounding medium. Consequently, this replacement can change the refractive index of the host material.

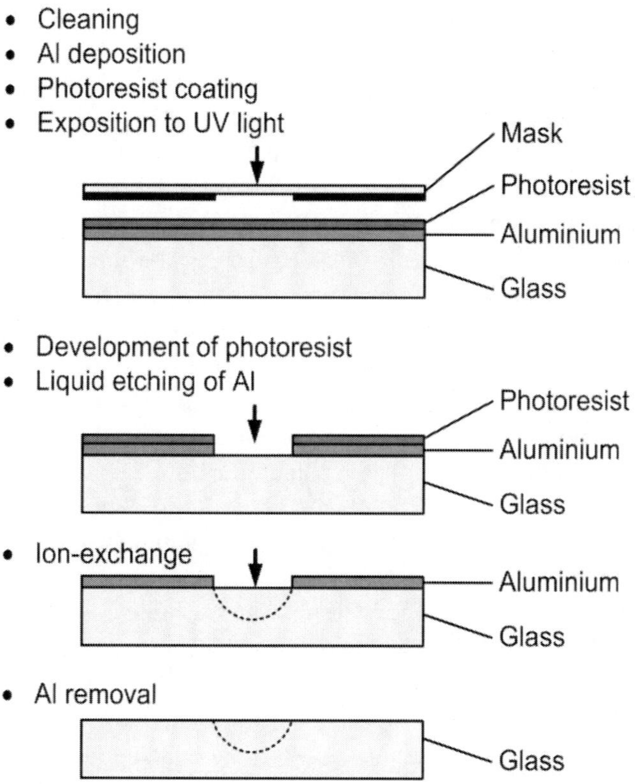


Figure 1. Step-by-step procedure for the channel waveguide fabrication by ion exchange [14].

The process to fabricate a channel optical waveguide in a planar substrate by ion exchange is shown schematically in Figure 1. After cleaning, an aluminum layer is deposited on the substrate and patterned with photolithography as the mask. Dielectric masks (usually a thermally oxidized metal film) can also be used when fabricating passive waveguides. The whole substrate is immersed into molten salt containing the ions to be introduced in the glass, in a temperature controlled furnace. After removing the residual salt and the mask, a bare glass surface with an ion-exchanged channel has been formed. Molten salt ion exchange is a purely thermal process. There is a difference of the concentrations of the two exchanged ions between the glass and the salt melt, and the mobility of ions in glass increases at an elevated temperature. This drives a balanced diffusion replacing the original ions in the glass by the ions from the melt. The change in glass composition locally increases the refractive index of substrate glass. Besides the material, the temperature and duration of the ion exchange are also important factors to influence the resulting sample. Diffuse nature of the process results in very smooth index profiles and extremely low-loss waveguides.

A convenient approach in ion exchange is to use a crucible furnace with a vertical chamber shown in Figure 2. A substrate holder is employed to control the insertion and removal of the substrate from the exchange bath while it reaches the desired temperature. For most purposes the furnace should be able to maintain and control the temperature with a range from 100 to 600°C, through utilizing the insulated cover and thermometers. The optional stirring bar is to make the melt more homogeneous.

Molten salt ion exchange is a purely thermal process. There is a difference of the concentrations of the two exchanged ions between the glass and the salt melt, and the mobility of ions in glass increases at an elevated temperature. This drives a balanced diffusion replacing the original ions in the glass by the ions from the melt. The change in glass composition locally increases the refractive index of substrate glass. There are several types of widely available glass substrates for ion exchange. The first and foremost one is the common microscope slide with high sodium content which can be easily replaced by ion exchange. For higher quality waveguides, some of the most commonly used glasses are borosilicates from Schott (BK7) and Corning (2947) because of the high sodium and potassium content and also the optical standard smoothness and flatness. Besides the material, the temperature and duration of the ion exchange are also important factors to influence the resulting sample. The typical multi-component glasses have a softening point near 700°C, and an annealing point in the vicinity of 550°C.

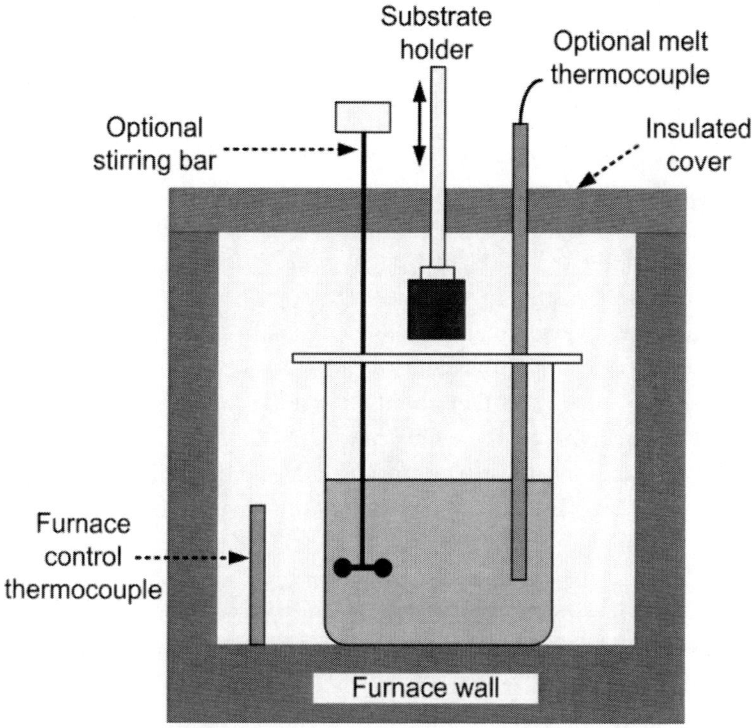


Figure 2. Furnace and an ion exchange apparatus cross section [14].

Based on the experience, the most reliable process temperatures are found at about 40 to 60°C above the melting point of the salt, to ensure high mobility of ions in the melt while avoiding problems associated with high temperatures. At the same time, it should be less than the temperature of salt decomposition. The duration is another factor influencing the refractive index change. The longer the duration is, the deeper the diffusion can reach. For fabricating a suitable waveguide, we should combine these two factors to make a balance. In the fabrication of optical channel waveguides, $Ag^+$-$Na^+$ ion exchange with a photolithographically patterned mask is a well-established process. When a conductive mask such as Al is used, one drawback is the optical absorption and scattering losses due to the deposited metallic Ag introduced during the ion exchange process. The initial idea to generate Ag nanoparticles in glass substrates stems from this observation. This technique with intentionally enhancing the formation of Ag nanoparticles is named as a masked ion exchange. When a conductive mask such as Al is used, one drawback is the optical absorption and scattering losses due to the deposited metallic Ag introduced during the ion exchange process. The initial idea to generate Ag nanoparticles in glass substrates stems from this observation. This technique with intentionally enhancing the formation of Ag nanoparticles is named as a masked ion exchange.

The masked ion exchange is ion exchange with a mask patterned photolithographically. The substrate used in this method is Corning 2947 glass. The masked ion exchange in silver-sodium salt melt is a well-established technique to fabricate optical waveguides [15, 16]. The diffuse nature of the process results in very smooth index profiles and extremely low-loss waveguides. However, metallic silver deposition at the edges of metallic masks is observed due to the existence of an Al mask [17]. The absorption and scattering losses of metallic silver could be avoided by gently etching the glass to remove the particles or by using nonmetallic masks such as aluminum oxide, silicon dioxide or silicon nitride [18], when making high-quality waveguides [19]. In this work, silver deposition in waveguide fabrication is used as an advantage, and it is used to fabricate Ag nanoparticles by intentionally enhancing the particle generation.

As shown in Figure 3, a standard photolithography process is applied to make the mask pattern. The masked glass wafers are immersed in the molten salt for 30min. For the characterization, Al mask is removed by phosphoric acid (H3PO4), and different thicknesses of glass layers are etched to expose the particles.

The standard photolithography process is shown schematically in Figure 3. by steps 1 and 2. Firstly, a 100-nm-thick Al film is deposited on the pre cleaned Corning 2947 glass by an electron beam evaporator. Next, the sample is baked at 120°C for 10 min to drive off the moisture and then primed in the adhesion promoter chemical hexamethyldisilazane (HDMS) to improve the adhesion of the photoresist to the wafer.

Secondly, a negative photoresist is coated by a spinner, with the speed of 6000 rpm for 20 s to produce a photoresist layer with a thickness of around 0.6 μm. A soft baking process at 90°C for 20 min is applied to drive off the excess photoresist solvent.

Thirdly, the sample is exposed under a UV light source through a prefabricated photomask for 30 s, to make the exposed photoresist soluble in the developer. After the sample is developed for 50s in the developer, a hard baking at 120°C for 10 min is done.

The exposure process is done by the mask aligner, and the baking process is done in an oven. By optical photolithography, different patterns have been done on the Al film.

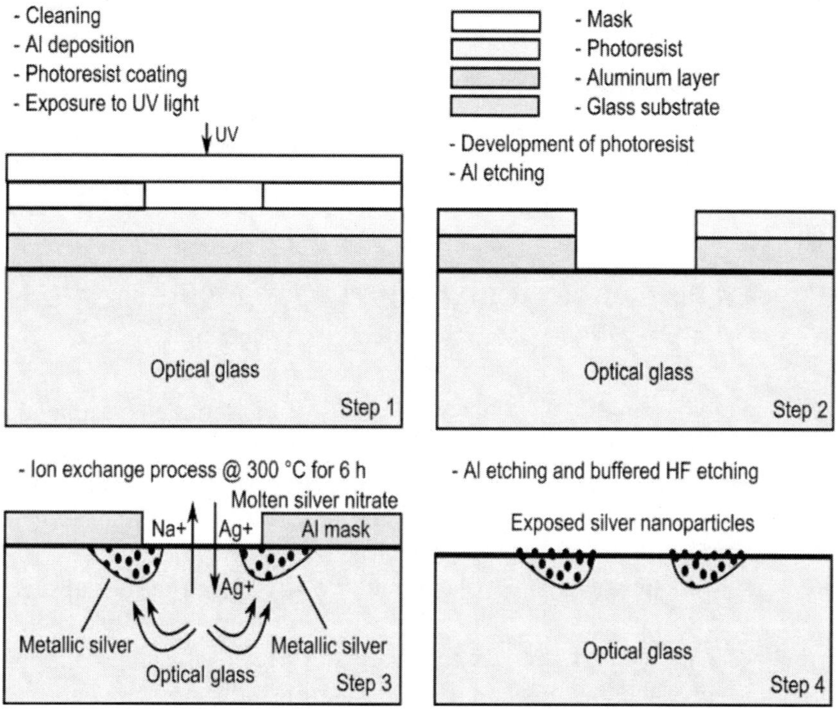


Figure 3. Masked ion exchange process to fabricate Ag nanoparticle pattern [20].

## 3. USING THE TWO-STEP METHOD IN ION EXCHANGE

In the two-step ion exchange, a Ag+-Na+ ion exchange described is utilized to exchange Ag+ ions into the Corning 2947 glass. After that, a 100-nm-thick Al film can be evaporated on ion-exchanged glass by using an electron beam evaporator method. Sample is dipped into a pure potassium salt (KNO3) at 400°C for 2 hours to promote the formation of Ag nanoparticles. The effects of Al film and KNO3 salt in the formation of particles can be study.

## 4. SETUP

The samples with very small features down to a micrometer and nanometer size are difficult to measure with a standard commercial spectrometer such as Perkin Elmer Lambda 950. Extraction of the influence by the small features from the large background imposes severe requirements on the instruments. Particularly, it is important to locate the position of the features. Therefore, a setup combining a microscope and a UV - VIS spectrometer is built to measure the absorption/transmission of Ag nanoparticle patterns with the diameter of 10 μm.

## 5. Ag Nanoparticle Substrates Fabricated by Two-Step Ion Exchange

After the Ag+ ion exchange for 6h at 300°C, if the sample is just annealed in the air for 2h at 400°C, the annealing will change the Ag+ concentration profile by lowering the peak of the Ag+ concentration and driving the whole profile deeper into the glass. However, the situation is different and more complicated in the two-step ion exchange method. The optical spectra reveal the important functions of both Al layer and the KNO3 salt in the subsequent process, and also give insight to the formation mechanism behind these two contributors. The Al film has an essential role during the galvanic replacement reaction, in which the $Ag^{+}$ ions are reduced into metallic Ag. In recent years, plasmonics as a subfield of nanophotonics has appeared and developed rapidly. It primarily focuses on the plasmon coupling effects within close proximity along metal nanostructures and on the influence of light at the nanoscale based on the properties of propagating and localized surface plasmons [1, 2]. Two promising applications for plasmonics include [21]: i) plasmonic waveguides to guide electromagnetic energy below the diffraction limit along for example continuous metal strips or the chains of closely spaced metal nanoparticles, and ii) plasmonic sensors to realize various types of sensing schemes based on surface plasmon resonances, such as refractiveindex plasmon sensing [4], surface enhanced Raman spectroscopy (SERS) [5], and metal enhanced fluorescence [22]. In the biosensing field of metal nanoparticles, one of the most important applications is SERS, because it significantly increases the practicability of many detection techniques by greatly enhancing the detected signal [23-26].

## CONCLUSION

Formation of the silver nanoparticles in waveguide fabrication with a masked ion exchange has been observed as a problem because it can increase attenuation in the waveguides. Here, the silver deposition is taken as an advantage, and it is used to fabricate Ag nanoparticles by intentionally enhancing the particle generation. The properties of these particles are studied and the formation mechanism of electrolytic deposition is discussed. Since the silver particles are only deposited at the edge of the mask, this method provides a way to make silver nanoparticle patterns. $Ag^+$ ions are introduced into glass in an $Ag^+$ as ion exchange, and can be reduced into the metallic silver in a subsequent $K^+$ ion exchange. Compared to silver particles formed by the masked ion exchange, the particles formed by the two-step ion exchange are over a large areas under Al film need to use high magnification. Moreover, the presented method could also enable the patterning of areas with the particles when applying a photolithographic mask in either the first step of Ag+ ion exchange or the second step of $K^+$ ion exchange process.

Ag nanoparticles embedded in glass can be prepared by a two-step method in with using the ion exchange Process. The particles can be formed by initialized the galvanic that can be replace by the reaction and the further enhanced by using the electrolytic deposition method. The mechanism is distinguishable from the metal particle formation [27, 28]. Potential differences between the mask, glass and the salt melt within distances can be in the micron range. The results demonstrate the formation of the particles and their SERS application.

## REFERENCES

[1] S. A. Maier, P. G. Kik, H. A. Atwater, S. Meltzer, E. Harel, B. E. Koel, et al., “Local detection of electromagnetic energy transport below the diffraction limit in metal nanoparticle plasmon waveguides,” *Nature materials,* vol. 2, pp. 229-232, 2003.

[2] E. Ozbay, “Plasmonics: merging photonics and electronics at nanoscale dimensions,” *Science,* vol. 311, pp. 189-193, 2006.

[3] J. F. Li, Y. F. Huang, Y. Ding, Z. L. Yang, S. B. Li, X. S. Zhou, et al., “Shell-isolated nanoparticle-enhanced Raman spectroscopy,” *Nature,* vol. 464, pp. 392-395, 2010.

[4] R. Kashyap and G. Nemova, "Surface plasmon resonance-based fiber and planar waveguide sensors," *Journal of Sensors,* vol. 2009, 2009.

[5] M. I. Stockman, "Nanoplasmonics: The physics behind the applications," *Physics Today,* vol. 64, pp. 39-44, 2011.

[6] A. J. Haes and R. P. Van Duyne, "A unified view of propagating and localized surface plasmon resonance biosensors," *Analytical and Bioanalytical Chemistry,* vol. 379, pp. 920-930, 2004.

[7] K. L. Kelly, E. Coronado, L. L. Zhao, and G. C. Schatz, "The optical properties of metal nanoparticles: the influence of size, shape, and dielectric environment," *The Journal of Physical Chemistry B,* vol. 107, pp. 668-677, 2003.

[8] M. Piliarik and J. Homola, "Surface plasmon resonance (SPR) sensors: approaching their limits?," *Optics express,* vol. 17, pp. 16505-16517, 2009.

[9] S. A. Maier and H. A. Atwater, "Plasmonics: Localization and guiding of electromagnetic energy in metal/dielectric structures," *Journal of Applied Physics,* vol. 98, pp. 011101-011101-10, 2005.

[10] D. Stuart, A. Haes, C. Yonzon, E. Hicks, and R. Van Duyne, "Biological applications of localised surface plasmonic phenomenae," in Nanobiotechnology, IEE Proceedings-, 2005, pp. 13-32.

[11] A. Y. Sonay, A. B. Çağlayan, and M. Çulha, "Synthesis of peptide mediated Au core–Ag shell nanoparticles as surface-enhanced Raman scattering labels," *Plasmonics,* vol. 7, pp. 77-86, 2012.

[12] T.-Y. Liu, K.-T. Tsai, H.-H. Wang, Y. Chen, Y.-H. Chen, Y.-C. Chao, et al., "Functionalized arrays of Raman-enhancing nanoparticles for capture and culture-free analysis of bacteria in human blood," *Nature Communications,* vol. 2, p. 538, 2011.

[13] J. R. Lakowicz, K. Ray, M. Chowdhury, H. Szmacinski, Y. Fu, J. Zhang, et al., "Plasmon-controlled fluorescence: a new paradigm in fluorescence spectroscopy," *Analyst,* vol. 133, pp. 1308-1346, 2008.

[14] S. I. Najafi, Introduction to glass integrated optics: Artech House Boston, 1992.

[15] S. Yliniemi, Q. Wang, J. Albert, and S. Honkanen, "Studies on passive and active silver–sodium ion-exchanged glass waveguides and devices," *Materials Science and Engineering: B,* vol. 149, pp. 152-158, 2008.

[16] A. Quaranta, E. Cattaruzza, and F. Gonella, "Modelling the ion exchange process in glass: phenomenological approaches and perspectives," *Materials Science and Engineering: B,* vol. 149, pp. 133-139, 2008.

[17] R. Walker, C. Wilkinson, and J. Wilkinson, "Integrated optical waveguiding structures made by silver ion-exchange in glass. 1: The propagation characteristics of stripe ion-exchanged waveguides; a theoretical and experimental investigation," *Applied Optics,* vol. 22, pp. 1923-1928, 1983.

[18] A. Tervonen, B. R. West, and S. Honkanen, "Ion-exchanged glass waveguide technology: a review," *Optical Engineering,* vol. 50, pp. 071107-071107-15, 2011.

[19] J. Zou and R. T. Chen, "Improvement of two-step K+–Na+ and Ag+–Na+ ion-exchanged glass waveguides by surface removal," *Applied physics letters,* vol. 88, pp. 081102-081102-3, 2006.

[20] Y. Chen, J. Jaakola, Y. Ge, A. Säynätjoki, A. Tervonen, S.-P. Hannula, *et al.*, "In situ fabrication of waveguide-compatible glass-embedded silver nanoparticle patterns by masked ion-exchange process," *Journal of Non-Crystalline Solids,* vol. 355, pp. 2224-2227, 11/1/2009.

[21] S. Lal, S. Link, and N. J. Halas, "Nano-optics from sensing to waveguiding," *Nature photonics,* vol. 1, pp. 641-648, 2007.

[22] J. R. Lakowicz, "Plasmonics in biology and plasmon-controlled fluorescence," *Plasmonics,* vol. 1, pp. 5-33, 2006.

[23] M. Moskovits, "Surface-enhanced Raman spectroscopy: a brief retrospective," *Journal of Raman Spectroscopy,* vol. 36, pp. 485-496, 2005.

[24] Z.-Q. Tian, "Surface-enhanced Raman spectroscopy: advancements and applications," *Journal of Raman Spectroscopy,* vol. 36, pp. 466-470, 2005.

[25] J. A. Dieringer, A. D. McFarland, N. C. Shah, D. A. Stuart, A. V. Whitney, C. R. Yonzon, et al., "Introductory LectureSurface enhanced Raman spectroscopy: new materials, concepts, characterization tools, and applications," *Faraday Discussions,* vol. 132, pp. 9-26, 2006.

[26] M. K. Hossain and Y. Ozaki, "Surface-enhanced Raman scattering: facts and inline trends," *Current science,* vol. 97, pp. 192-201, 2009.

[27] Y. Chen, J. Jaakola, Y. Ge, A. Säynätjoki, A. Tervonen, S.-P. Hannula, et al., "In situ fabrication of waveguide-compatible glass-embedded silver nanoparticle patterns by masked ion-exchange process," *Journal of Non-Crystalline Solids,* vol. 355, pp. 2224-2227, 2009.

[28] Y. Chen, J. J. Jaakola, A. Säynätjoki, A. Tervonen, and S. Honkanen, "Glass-embedded silver nanoparticle patterns by masked ion-exchange process for surface-enhanced Raman scattering," *Journal of Raman Spectroscopy,* vol. 42, pp. 936-940, 2011.

In: Systems Engineering
Editor: Marlene Hopkins
ISBN: 978-1-63485-752-9


*Chapter 5*

# SYSML: THE DIALECT FOR MODEL-BASED SYSTEMS ENGINEERING

***Ana Luísa Ramos*[*] *and José Vasconcelos Ferreira*[†]**
DEGEIT - Department of Economics, Management, Industrial Engineering and Tourism,
GOVCOPP Research Unit, University of Aveiro, Aveiro, Portugal
Campo de Santiago, Aveiro, Portugal

## ABSTRACT

A successful Model-Based Systems Engineering (MBSE) environment aims to construct a model of a system that can be transformed into the "real thing" through a given MBSE methodology. Modelling is critical to achieve the overall "maximum" performance of the system through a cooperative, integrative, adaptable and interoperable environment. So, the main artifact of a MBSE methodology is the System Model.

Model the system (or Design the system) is a central piece of the Systems Engineering' process and results in a complete low-level description of the system-of-interest that will guide the successful development and deployment of that system. This success is measured by the fulfilment of the stakeholders' expectations and, consequently, by their satisfaction.

[*] Corresponding author e-mail: aramos@ua.pt.
[†] E-mail: josev@ua.pt.

The effective utilization of graphical modelling languages able to support collaborative development environments and successful stakeholders' communication and interactions constitutes a major issue when developing the System Model. SysML (Systems Modeling Language) is a UML profile or dialect for Systems Engineering and is currently the state-of-the-art modeling language for MBSE enhancing the development of successful systems in collaborative environments.

The chapter highlights the major advantages of using graphical modelling languages in a MBSE context and describes the main features of SysML, the modeling language that the authors believe to be "the dialect" for MBSE. This description is illustrated with original SysML diagrams related to the human body (probably, one of the most complex systems in our world).

**Keywords**: collaborative development, MBSE, system model, SysML, systems engineering

## INTRODUCTION

Systems Engineering (SE) aims to tackle the complex and interdisciplinary whole of socio-technical systems made of other systems, personnel, hardware, software, information, processes, and facilities, thereby providing the means to enable their successful realization. As stated in the Systems Engineering Body of Knowledge (BKCASE Editorial Board, 2016), "Systems Engineering focuses on ensuring the pieces work together to achieve the objectives of the whole". An integrated holistic approach is crucial to develop these modern systems and take proper account of their multifaceted nature and numerous interrelationships. As the system's complexity and extent grow, the number of parties involved (i.e., stakeholders and shareholders) usually also raises, thereby bringing a considerable amount of points of view, skills, responsibilities, and interests to the interaction. In addition, several challenges must be addressed: join both technological and social perspectives, include customized services with complex human-centred aspects (Tien, 2008), incorporating an extensive set of " ilities" like flexibility, sustainability, real time capability, adaptability, expandability, reliability, usability, and delivery of value to society (Rhodes, 2008).

Modelling these systems is critical to achieve their overall "maximum" performance through cooperative, integrative, adaptable and interoperable environments. Model-Based Systems Engineering (MBSE) can be described as the formalized application of modelling principles, methods, languages and tools to the entire life cycle of large, complex, interdisciplinary, socio-technical systems. MBSE approaches have the potential to vastly improve the systems engineering of modern, complex systems using digital models to describe and represent all aspects of a system (Bjorkman et al., 2013). This model-centric approach, which main artefact is a coherent model of the system being developed, contrasts with the traditional document-based one (Friedenthal et al., 2008).

The major potential advantages of this emergent paradigm include, for instance, enhanced communications, shared understanding and knowledge capture, improved design precision and integrity, better development traceability, and reduced development risks. In the next decade, it is expected that MBSE will play an increasing role in the practice of Systems Engineering and will extend its application domains beyond hardware and software systems, including social, economic, environmental, and human performance dimensions.

The Systems Modeling Language (SysML) is a UML profile devoted to Systems Engineering. This graphical language for modeling systems, which supports the specification, analysis, design, and verification of complex systems, is considered as the next *de facto* modeling language for SE and MBSE (Ramos et al., 2013). The transition from a document-based paradigm to a model-based paradigm is incremental and evolves as the methods and tools mature increasing the promise of MBSE (BKCASE Editorial Board, 2016). As in other fields of knowledge (e.g., software development), this modeling language needs to become a standard regarding the widespread adoption of MBSE within organizations. This standardization is crucial to advance the field and to establish benchmark practices across different domains.

This chapter highlights some modeling fundamentals and the major advantages of using graphical modelling languages in a MBSE context. It also describes the main features of SysML, the modeling language that the authors believe to be "the dialect" for MBSE, using original SysML diagrams.

## MODELING FUDAMENTALS

Modelling is a typical task of the scientific activity. The generated models are undoubtedly a main instrument of modern science. A model (the term "model" derives from the Latin word *modulus*, which means measure, rule, pattern, example to be followed (Ludewig, 2003)) is a representation of a selected part of the world (with actual existence or projected) which constitutes the domain-of-interest. That representation is an abstract view of the thing being modelled in order to capture the important aspects, from a certain point of view, simplifying or omitting the irrelevant features (Rumbaugh et al., 1999). Ludewig (2003) describes three criteria that a model must meet to be elected as a model: mapping criterion (there is an original object or phenomenon that is mapped to the model), reduction criterion (not all the properties of the original are mapped onto the model and this one must mirror at least some properties of the original), and pragmatic criterion (the model is useful that is, it can replace the original for some purpose).

According to Rumbaugh and colleagues (1999), the models are important to: i) capture and state requirements and domain knowledge so that all stakeholders may understand them; ii) think about the design of a system; iii) produce usable work products; iv) organize, find, examine, filter, manipulate, and edit information about large systems; v) explore several solutions operationally, economically, and environmentally; and vi) master complex systems. Sussman (2000) adds the importance of using models to gain insight into complex systems, to do experimentation (in a considerable number of cases is impractical, due to cost, technical or other constraints, to experiment on the real systems and to get direct measurements), to operate systems in real-time (real-time modelling), and to negotiate, with conflicting parties, how the system will be deployed. He synthesizes the models' functions as following: "we model to understand, we model to explain, we model to predict, and we model to improve". Buede (2009) states that the main purpose of a model is to answer a set of questions that we could not answer without it and, additionally, to understand why the answers are what they are that is, to gain insight into how the world functions.

The underlying shared idea is to comprehend the reality in analysis through some kind of simplification depending on the purposes of the study and the model. According to the Stanford Encyclopaedia of Philosophy, the models are vehicles to explore, to understand, and to learn about the world, being this cognitive function the basis of the so-called 'model-based reasoning'. Learning occurs with denotation (defining a representation relation between the model and the target), demonstration (investigating the characteristics of the model in order to demonstrate theoretical conjectures), and interpretation (converting the findings into claims about the target system) (Hughes, 1997). These activities require a deep analysis of the system to be modelled enhancing its understanding.

The modeller tends to shape his view of the system according to his favourite(s) modelling approach(es) like, for example, simulation, network analysis, system dynamics, control theory, cost/benefit analysis. The "art" of choosing the best approach and representing the model adequately, in order to answer to the target questions, constitutes the major characteristic of a good modeller. Frequently, those decisions are closely related with time and budget constraints, and the availability of data. The creation of a good model involves the definition of the questions the model should be able to answer, the development of the model, the verification, the refinement, the validation, and the corresponding evaluation. The expertise on the selected approach(es) is crucial to develop the model adequately and to verify it (ensuring that the model was developed correctly). The validation of the model guarantees that the right model has been built. This validation can have several dimensions namely, the conceptual validity which addresses the structure of the model and the assumptions made, the operational validity which looks into the behaviour of the model's outputs ensuring a valid representation of the real target system, and the data validity that addresses whether the appropriate inputs (individuals, observations, statistical distributions, etc.) were used in building, testing, and using the model (Law, 2014; Buede, 2009).

The success of the model is then measured by their users in different ways and according to their perspective/expectations of/on the model purpose. Criteria such as reliability, completeness, accuracy, power to convince, ease of use, compatibility, run time, and extendibility are of frequent utilization. According to Karcanias (2004) modelling is "the common basis to human activities and thus its development is also a measure of our ability to understand nature, society, and related issues."

There are several model taxonomies provided by philosophers, biologists, mathematicians, engineers, economists, software developers, etc. Their model-types are usually related with the nature of the systems under study and with the questions to answer. The general classification proposed by the Stanford Encyclopedia of Philosophy includes a categorization in terms of what they represent. Models can represent phenomena, data, and theory. The first category includes scale models which are materialized into physical objects (down-sized or enlarged copies of the target systems; naturalistic replicas such as model bridges, or wind tunnels), idealized models which are represented by fictional objects (representations made through the isolation of some remarkable characteristics of a system and their distortion into an extreme case, such as the infinite velocities, isolated systems, the planetary system mechanic' model, or markets in perfect equilibrium), analogical models which are usually represented by descriptions (representations which recur to similar systems, based on similar structure or properties, such as the billiard model of a gas, or the computer model of the mind), and phenomenological models (representations that reflect only the observable properties of their targets, such as the liquid drop model of the atomic nucleus). The data models usually entail the reduction of data and curve fitting, requiring sophisticated statistical techniques (e.g., the model of a trajectory of a planet). The models representing theory are structures that make all sentences of a theory true (for example, the finite state machine model representing the theory of finite state machines).

A simpler taxonomy is presented by Whitten and colleagues (2004), within the Systems Analysis and Design area, and includes the distinction between physical and logical models. The first category also known as implementation or technical models depicts what the system does and how the system is physically implemented. The logical models also known as essential, conceptual or business models are pictorial representations of what a system is or does.

Sommerville (2010) presents several model classifications in his book Software Engineering. He considers three types of models for describing the software process namely, workflow models, dataflow or activity models, and role/action models, which are based on three general development models: the waterfall, the iterative incremental or evolutionary, and the component-based. When describing the systems' models developed during the requirements engineering process, the author enumerates four major types of models regarding the representation of the system from different perspectives namely:

i. the external perspective which leads to context models that describe the context, the boundaries, and the environment of the system to be modelled;
ii. the behavioural perspective that is represented by behavioural models that describe the actions of the system;
iii. the structural perspective that is illustrated by data models which express the architecture of the system and the logical form/structure of the data to be processed;
iv. the object-oriented perspective, that combines behavioural and structural modelling and is represented by object models that reflect the real world entities that are manipulated by the system.

The explanation of the architecture of the system also includes a categorization of models: static structural models, dynamic process models, interface models, relationship models, and distribution models, as well as the organization of the system: repository models, client-server models, and layered models. In the requirements engineering tasks and design of user interfaces the author suggests the utilization of prototype models that can be throw-away prototypes (for example, paper prototypes, storyboards, 'wizard of Oz' prototypes) or evolutionary prototypes.

Sánchez and colleagues (2009), studying the area of information systems, propose a main division consisting of models as originals (a reality serving as example) and models as copies (a representation of a domain according to a specific point of view). The former category includes the metamodels ("models as originals to describe the models as copies"), the architectural styles ("abstractions of types of elements and formal aspects from different specific architectures"), the patterns ("descriptions of communicating objects and classes that are customized to solve a general design problem in a particular context"), and the process models ("abstract representations of a process from some particular perspective"), whereas the latter category embraces the schemas ("abstract and organized representations of a domain"), diagrams ("abstract and organized graphic representations of a domain" or graphical schemas), ontologies ("explicit specifications of a conceptualization" that is, the process of creating a concept of something), and architectures ("the fundamental organization of a system embodied in its components, their relationships to each other and to the environment, and the principles guiding its design and evolution").

As obvious, the presented taxonomies, as well as a considerable additional number not mentioned in this chapter, are diverse and are intimately related with the reality to represent. It is also common to found different names meaning the same type of abstraction.

## Modeling in MBSE

Model-Based Systems Engineering (MBSE) can be formally stated as "the formalized application of modelling to support systems requirements, design, analysis, verification, and validation activities beginning in the conceptual design phase and continuing throughout development and later life cycle stages" (BKCASE Editorial Board, 2016).

The main principle underlying this approach relies on the creation of a coherent model of the system being developed. This model centric approach is expected to replace, in the next years, the traditional document centric approach that is based on documents written in text.

Modelling in Systems Engineering/MBSE means primarily to create a shared vision among the system's stakeholders, to specify, describe, communicate, and test that shared vision, to estimate or predict some quantitative measure of the system, and to select design options (Buede, 2009; Van Daalen et al., 2009). As Buede (2009) also states "In fact, models are so pervasive in the engineering of systems that engineers must always remind themselves not to confuse reality with the models of reality that are being created, tested, and used." The referred author proposes a classification, more related with the field of SE, involving a high-level classification based on four major categories namely, physical, quantitative, qualitative, and mental models.

The *physical or scale models* are three-dimensional representations of the target system and can be of type: full-scale mock-up (used to enable the visualization of the physical placement of elements of the system, or to match the interfaces between systems and components), subscale mock-up (used to examine a certain issue), electronic mock-up (used to test complex arrangements of electronic components), and breadboard (a board to build and test electronic or mechanical prototypes).

The *quantitative or mathematical models* endow with numerical answers and are subdivided in analytic, simulation, and judgemental models. These subcategories can be further divided into deterministic/stochastic, discrete/continuous, and static/dynamic. The analytic models are based on a set of equations that can be solved to find a set of solutions, the simulation models are used when the analytical methods are not realistic and there is a need to test complex operational scenarios, and the judgemental models provide representations based on opinions of experts, which are particular useful when there is lack of information or as support for subsequent modelling activities.

The *qualitative models* make available symbolic, textual or graphical answers. The symbolic models are based on logic or set theory. The textual models provide verbal descriptions. The graphical models, also referred as visual models, use elements of mathematical graph theory or simply artistic graphics to represent structure, flow, behaviour, etc.

The *mental models*, which can be considered as the basis of all the other types of models, are individualized mental representations of the problem and the solutions. Prior to the development of physical, quantitative, and qualitative models every engineer or modeller has his own mental model that needs to be "translated" into unambiguous representations that can be clearly shared and discussed with other parties.

The present chapter emphasizes the modelling approach as a means to objectively create, specify, describe, communicate, and test a shared vision among the stakeholders of the system. In that sense, the category of qualitative models will be the elected group. In this group, the graphical models have always had a particular role in the MBSE field. The pictorial representation and the amount of information that it can handle, as well as the facility to be stored in our memory, is frequently illustrated by the aphorism "A picture is worth a thousand words". It means that long complex verbal descriptions (processed by the left brain) can be synthesized in a visual representation (processed by the right brain) that is most effectively in the communication and understanding of the verbal message. This kind of representation enables communications and displays information at different levels of abstraction, helping to manage complexity. As Peter Senge (1990) states "If we want to see system wide, we need a language of interrelationships [R-mode]". The "ultimate" performance can be achieved with the inter-play between the left and the right brain or the whole-brain thinking, which allows mixing science and art, premises and creativity, words and pictures.

The modelling techniques used in the field of SE/MBSE, to develop systems (modelling concepts, properties, attributes, structure, behaviour, entities, interactions, relations, etc.), have always been predominantly qualitative and based on a graphical or pictorial representation. These techniques require a corresponding describing language (graphical modelling language or visual modelling language), used to represent reality, that involves semantics (set of symbols or signs which form the basis of representations) and syntax (the proper ways of combining the symbols and signs to form thoughts and concepts).

The first representations, with Functional Flow Block Diagrams (FFBD) for ballistic missile development, date back to the 1950s and have been a classical modelling approach within the SE community for long decades. In 1975, J. Long had extended these diagrams, adding an input/output element, and creating a behaviour diagram that depicted the sequential relationship of the functions (blocks) performed by the system with a "control" flow (Oliver et al., 2009). By the time it was also published a methodology, the Structured Analysis and Design Technique (SADT™), borrowed from the work of Long, which described a system as a hierarchy of functions (Ross, 1977). This technique, developed by software engineers at IBM (Oliver et al., 2009), used a top-down decomposition approach but did not represent the interactions between the parts and the emergent properties of the whole.

In the 1990s the NIST launched the IDEF0 (Integration DEFinition for Functional Modeling), a part of the IDEF suite of modelling tools derived from the SADT. This graphical language is mainly devoted to the representation of activities or processes illustrating the functional perspective of the system, as well as the data flow and the system control. The enhanced FFBD and the IDEF0 have been the main modelling languages used in SE in the last decades. Other tools include, for example, the N-Squared Charts, State-Transition Diagrams, and Petri Nets.

The modern object-oriented practices, with its roots in Software Engineering, are now pervasive in the Systems Engineering/MBSE field. Oliver and colleagues (2009) trace the origin of the object-oriented paradigm back to the 1970s, with the development of abstract data types and the introduction of classes to programming languages, like Simula67, in order to provide procedure, data, and control abstractions. In the 1990s these principles have been extended to the analysis and design of software, through the Booch method, the Object Modelling Technique (OMT) (Rumbaugh et al., 1991) and then through the de facto Unified Modeling Language (UML) (Booch et al., 1999). The gurus Booch, Jacobson, and Rumbaugh, are the main authors of this universal graphical modelling language for software development that is now under the responsibility of the Object Management Group (OMG). The characteristics of the software systems are different from those of the systems of SE (that may also include software components) and consequently, the UML lacks support from aspects like the whole/part decomposition, the interconnections provided by physical things (and not by compilers), or the trade studies. In order to incorporate these and other features, the OMG and the INCOSE have joined efforts and developed an extension of UML for Systems Engineering, the Systems Modeling Language (SysML) (Friedenthal et al., 2008), that has been released in 2007. The OMG SysML Version 1.4 specification has been formally released in 2015. According to Oliver et al. (2009) SysML has its limitations but has extended some concepts in "useful ways beyond historic systems engineering practice."

The Object-Process Methodology (OPM), founded by Dov Dori in 2002, and the corresponding graphical and textual representations, OPDs and OPL, enlarge the domain of object-oriented modelling tools for SE. According to its author, OPM "is a comprehensive novel approach to systems engineering. Integrating function, structure and behaviour in a single, unifying model, OPM significantly extends the system modelling capabilities of current object-oriented methods" (Dori, 2002). In fact, SysML and OPDs/OPL constitute the current state-of-the-art systems modelling languages. Being SysML a more "institutionalized/standardized" language with the support of the OMG and the INCOSE, and the OPDs/OPL a more intuitive simpler language with less training effort, it seems interesting to combine the advantages of both languages creating synergies between them (Grobshtein and Dori, 2009). This integration can strongly contribute to a common understanding of the system and to improved communications between different stakeholders, as well as to a proficient Systems Engineering collaborative development environment (Gomes et al., 2011).

Due to its support by OMG/INCOSE, availability in commercial tools, and widespread adoption in industrial and defense sectors, the authors believe that SysML will be the elected graphical language for MBSE. In fact, as described by Durugbo (2013), "evidence exists to suggest that research involving the use of SysML has continued to increase over the years" showing that "SysML research is relevant, timely, and flourishing."

## SysML: The Dialect for Model-Based Systems Engineering

As previously mentioned, the Object Management Group Systems Modeling Language (OMG SysML™) is the UML profile or dialect for Systems Engineering. This graphical modelling language supports the analysis, specification, design, verification, and validation of complex systems that may include hardware, software, data, personnel, procedures, facilities, and other elements of man-made systems (Friedenthal et al., 2008). SysML represents a subset of UML 2 with extensions to cope with the particularities of systems modelling namely, the issues of requirements modelling, interdisciplinarity, hierarchical decomposition, quantitative analysis, and trade studies. According to David et al., (2010), this language provides a comprehensive system specification paradigm and avoids the software vocabulary of UML.

Moreover, SysML is having the same success in the field of Systems Engineering as UML has in the software development industry.

The language results from conjoint efforts of OMG and INCOSE, through the Systems Engineering Domain Special Interest Group, with a request for proposal in 2003 containing the requirements for a systems modeling language aligned with the AP233, in order to support tool interoperability (and integration with other engineering models). The SysML Partners (tools vendors, leading industry users, academia, professional organizations, and government representatives) (OMG SysML, 2010) answered to this request and submitted a first version, to the OMG, in 2005. The final version was adopted by the OMG as OMG SysML™ in 2006, and the first official specification (version 1.0) was released in 2007.

The current available specification (version 1.4) was releases in 2015 and the modeling language is now implemented in several commercial tools such as Artisan Studio (from Artisan Software) and Rhapsody (from IBM/Telelogic), is described in diverse books and papers, and is offered in sevral academic courses. Regardless this pervasive knowledge, the language is still young claiming for a wide adoption and further real experiences.

The SysML metamodel (the abstract syntax that describes the language concepts, characteristics, and interrelationships) is, like UML, described by the OMG Meta Object Facility (MOF). The SysML model (or user model) is a description of a system and its environment for a definite purpose and includes model elements, which have a concrete syntax that must conform to the language metamodel. In a profile, such as the case of SysML, some metaclasses are replaced by stereotypes that create new or modified concepts.

The SysML modelling tools usually store the user model as structured data in a model repository and the model enters and retrieves that information by using the graphical representation (i.e., the diagrams). The SysML diagrams, which reflect various aspects of a system, are nine and are organized in four major blocks that are known as the four pillars of SysML and represent four key modelling facets: the requirements of the system, the structure, the behaviour, and the parametric relationships. These different views match particular viewpoints (the stakeholders' perspectives) and enable the holistic approach required by Systems Engineering/MBSE.

SysML reduces UML size (both in the number of diagrams and in the number of total constructs) and the related software bias, and adds two critical diagrams for Systems Engineering activities, the requirements and the parametric ones (Figure 1). These are essential to support requirements engineering and performance analysis.

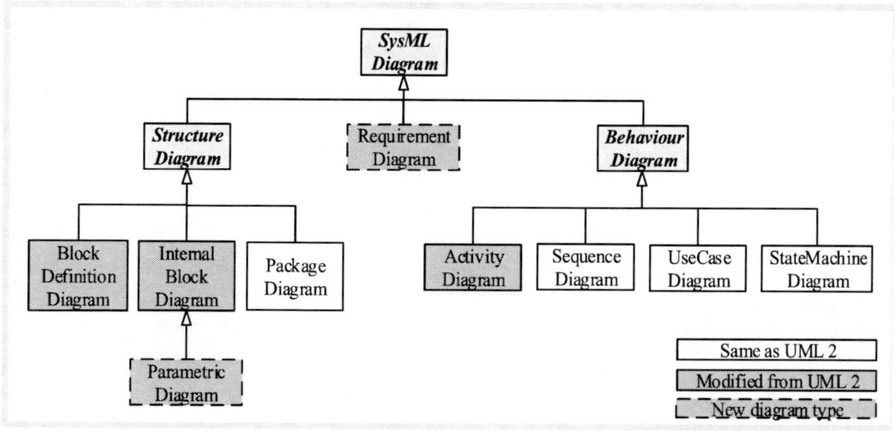


Figure 1. SysML diagram taxonomy (adapted from OMG SysML, 2016).

The SysML basic unit of structure is the block that can represent hardware, software, facilities, personnel, or other element of a system. The structure of the system is described by Block Definition diagrams (bdd) and Internal Block diagrams (ibd). The system's model is organized through the Package diagram (pkg). The Parametric diagram (par) depicts constraints on property values such as performance, acceleration, reliability, and is particularly useful to integrate the system's specification and design models with other engineering models. The behaviour of the system is described through the Use Case diagram (uc) that depicts the system's high-level functionality, the Activity diagram (act) that represents the flow of data and control between activities, the Sequence diagram (sd) that represents the interaction between collaborating parts of a system, and the Sate Machine diagram (smd) that describes the state transitions and actions that a system or its parts perform in response to events.

The novel Requirements diagram (req) allows the hierarchical representation of text-based requirements and the "satisfy" and "verify" relationships enable the modeller to relate the requirement to a model element. This diagram fills the gap between the typical requirements management tools and the system model (OMG SysML, 2016).

The language also provides non-graphical representations in the forms of tables, matrices, and trees. One of the most common tabular representations are the allocation tables which are dynamically derived tables from the SysML "allocation" relationships and can represent the allocation of functions to components, logical to physical components, and software to hardware. Each

SysML diagram represents a model element and has a frame enclosing a contents area (canvas) where the diagram is designed, and a left header that provides the diagram context in the following manner: diagramKind [modelElementType] modelElementName [diagramName]. For example, the header bdd [Package] Structure [Structure Hierarchy] refers to a block definition diagram called Structure Hierarchy that represents a package element named Structure. Additionally, the diagram can have a textual description attached to the frame detailing the version, status, and purpose of the diagram.

The following paragraphs provide a brief description of each kind of diagram and the most relevant associated features. The description is based on some of the major literature references: Wilkiens (2007), Friedenthal et al. (2008), and Friedenthal et al., (2009). The description is illustrated with original SysML diagrams which are related to the human body system (probably one of the world's most complex systems but familiar to everyone) and were developed in the tool Artisan Studio®.

## Package Diagram (pkg)

The organization is essential to manage a SysML model, facilitating reuse of model elements, easy access and navigability, configuration management, and data exchange with other tools. This diagram (pkg) is used to organize the model, being a model organized into a single hierarchical structure of packages, similar to the folders' arrangement in a Windows directory structure. Each package, the fundamental unit of organization, contains a set of model elements which are logically grouped packageable elements such as blocks, use cases, activities, value types, and also packages (nested packages) (Figure 2b). The criteria to organize the model are usually related with the chosen methodology but the organization by system hierarchy, the organization by diagram type, and the organization by working teams are typical examples of arrangement principles. The package is also a namespace for its members, defining a set of rules for their unique identification (each element of a particular element type within the package must have a unique name). The header for a package diagram is as follows: pkg [package type] package name [diagram name]. The package type can be a model, a package, a model library, a view or a profile. A model is a special kind of high level package that contains a set of model elements that describe a domain of interest (Figure 2a). A model library is a package that contains reusable model elements for a given domain (Figure 2c). A view is a type of package that conforms to a viewpoint which, in turns, describes a perspective of interest to a group of stakeholders that is used to specify a view of a model (Figure 2d). A profile is a type of package to group a set of stereotypes (language extension mechanisms) and supporting definitions that are related to the domain. The viewpoint includes a collection of properties that identify the purpose for taking the perspective, the stakeholders who have interest in the perspective, the concerns that the stakeholders want to address, the languages used to present the view, and the methods used to establish the view. The view provides the model information that addresses the stakeholders concerns.

The Figure 2 depicts examples of pkg that represent some of the mentioned aspects. The diagram a) shows the top level packages within the model of the human body (organized by the major approaches used in anatomy studies) including a package Value Types that uses the imported definitions of units and dimensions from the reusable library package SI Definitions (from the SysML profile). The diagram b) represents the Organ Systems package with hierarchically nested packages for the Cardiovascular one. The diagram c) details the Internal Organs package showing some of its elements (blocks) that can be (re)used in the Organ Systems, Surface Anatomy, etc. The diagram d) illustrates the Cardiac Contraction view conforms to the Cardiac Mechanism Viewpoint which highlights the mechanical performance of the heart. The view, especially important for the patient, the cardiologist, and the medical technician, imports packages that contain elements related with the cardiac mechanism.

## Block Definition Diagram (bdd) and Internal Block Diagram (ibd)

These diagrams are devoted to the representation of the system's structure in terms of its hierarchy and interconnections. The bdd is used to capture block properties and relationships such as composition, association, specialization and can represents a package, a model, a model library, a block or a constraint block. The ibd is mainly used to define the internal structure of a block, how its internal parts are connected and what flows between them, and always represents a block.

The block is the fundamental SysML unit of structure (like the UML class) that can abstract any level of the system hierarchy (the top level system, a subsystem, logical or physical components). Blocks describe a system as a collection of parts each playing a certain role in a defined context, and are a description of a set of similar instances. The block possesses information about its static structure, its behaviour (how the block responds to a stimuli), and its constraints that can be described in a series of standard compartments labelled by a proper heading like parts, references, values, operations, reception, constraints. The name compartment is mandatory (with the optional keyword "block") and is at the top of the rectangle (Figure 3). The parts, references, and values are the properties of the block which describe its structural aspects. The part properties (also blocks) describe the hierarchical decomposition of an enclosing block, defining the part in the context of its whole. The part is the usage of a block in the context of a composing block (also known as role). The reference properties describe weaker relationships than the composition one, such as associations or aggregations, and represent a hierarchy that references blocks that are part of other composition hierarchies or stored items. The value properties, which are defined by value types, describe quantifiable physical, performance, and other characteristics such as weight or volume.

The ibd represents the internal structure of a composite block that is designated by the diagram frame in the block name, providing the context for all the model elements on the diagram (Figure 4). The parts of the composite block can be connected providing an interaction specified by the parts' behaviour. The ports are used to connect the parts of a block and describe the block's points of interaction (on its boundary) that is, the block's interface. They can be standard ports (specify the types of services that a block requires or provides) or flow ports (specify what can flow in and out of blocks). The selection of the type of port to use is a methodological choice that is typically related with how the behaviour is expressed. Usually, flow ports represent continuous flows of physical entities or other continuous or discrete flows sent from one process to another. The standard ports are more appropriate to describe a system whose behaviour is represented by the invocation of services (e.g., interfaces between software components). An item is used to represent an entity that flows through a system. These entities (physical or information) can be modelled as blocks, value types or signals. The items can be defined at different levels of abstraction throughout the design process. An item flow specifies the items that flow across a connector in a particular context. Item flows on connectors between ports must be compatible with the port definitions but may be more specific (e.g., a flow port may be typed fluid but, the specific fluid for a given context may be typed blood).

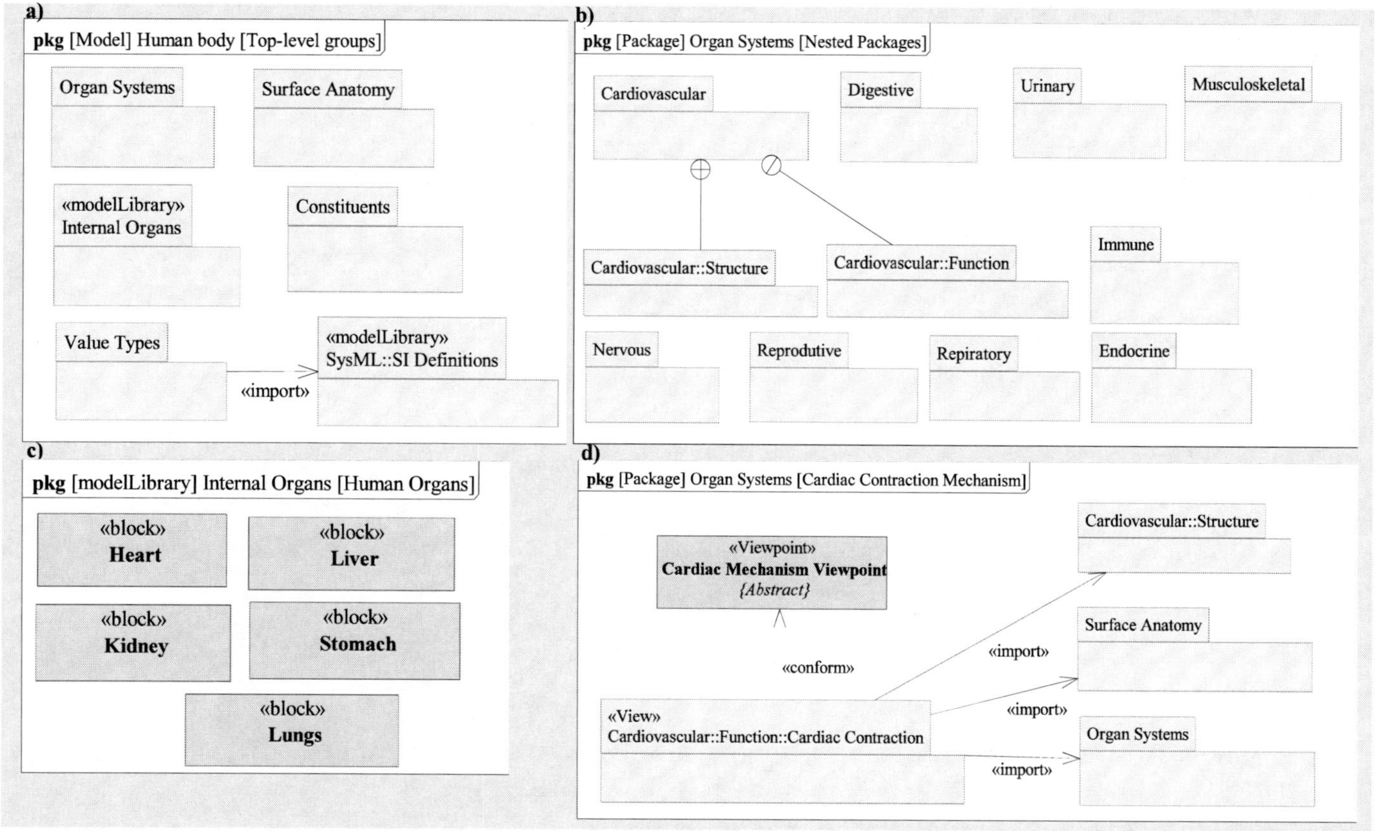


Figure 2. Examples of SysML package diagrams.

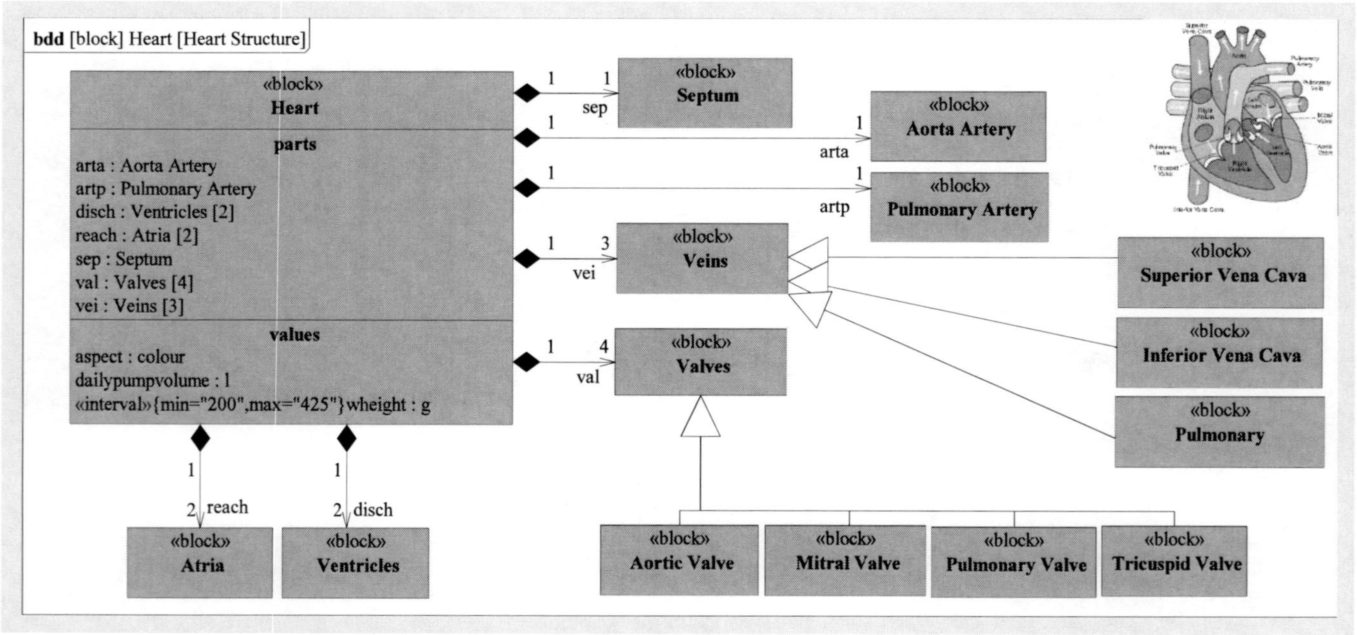


Figure 3. bdd for a portion of the composition hierarchy of a human heart (like an assembly relationship) showing some properties (parts and values) of the heart block and some generalization relationships.

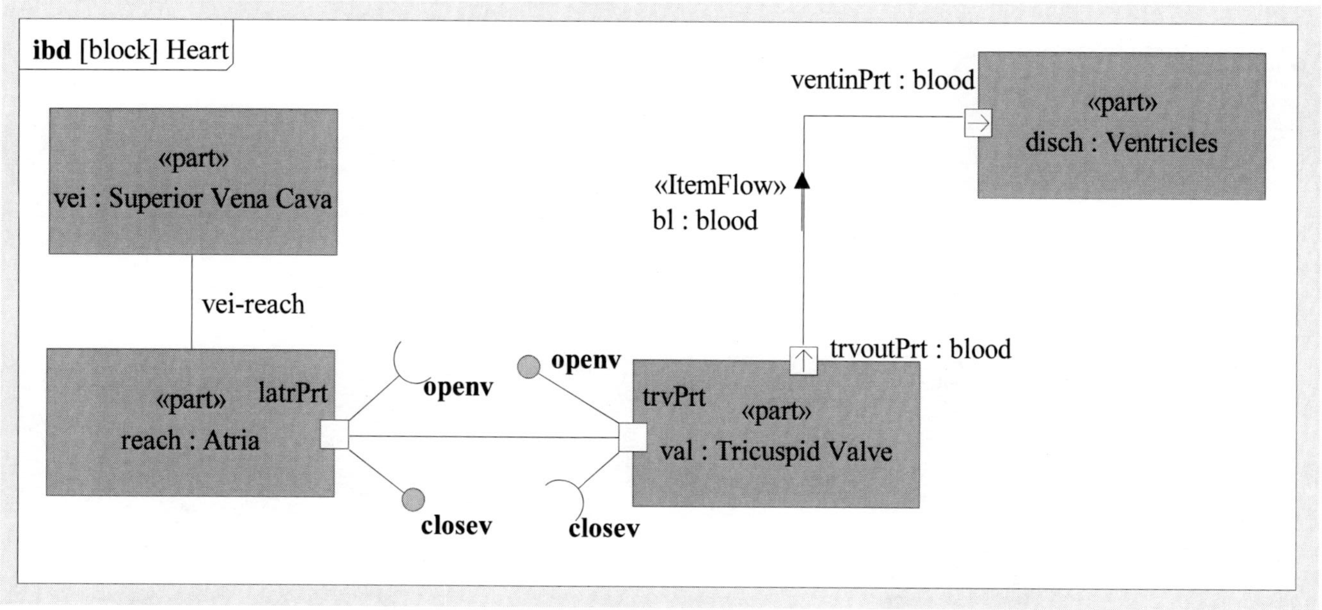


Figure 4. ibd for a portion of the internal structure of the composite block Heart depicting an assembly connector vei reach, a standard port latrPrt with a required interface openv and a provided interface closev, and an atomic flow port with an item flow named blood.

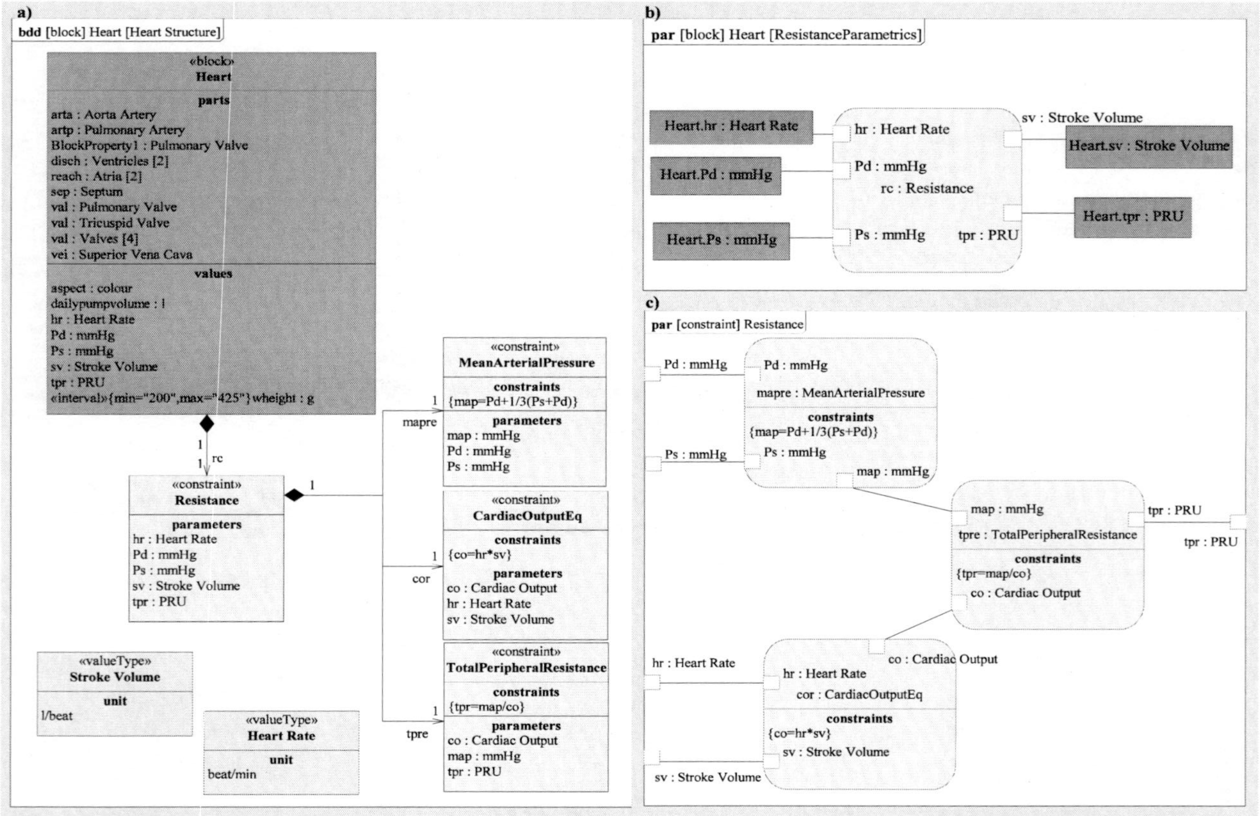


Figure 5. Parametrics modeling: a) bdd for the constraint block ‘Resistance’ that constraints the value properties of the block ‘Heart’ and is defined by composition of other constraint blocks, b) par for the block ‘Heart’, c) par for the constraint block ‘Resistance.’

## Parametric Diagram (par)

The parametric diagram (a novelty of SysML) enables the modelling of constraints on the physical and performance properties of the system providing support for additional engineering analyses such as trade off studies and sensitivity analysis. The par can represents a block (Figure 5 b) or a constraint block (Figure 5 c). The constraint block is a special kind of reusable block that has the keyword "constraint" and two main features: a set of parameters (scalars, vectors or structured data types that may have units, dimensions, probability distributions and are defined as follows: parameter name: type[multiplicity]) and a mathematical or logical expression (that can be time dependent) that constraints the parameters. The constraint block can be defined using other constraint blocks through composition. The value properties of a block can be constrained on a bdd by representing composite associations between the block and the required constraint blocks (Figure 5 a). The definition of a constraint block is represented on a bdd and its use on a particular context, called constraint property, is depicted on a parametric diagram. A constraint can also be represented in the constraints department of a block or with an attached note to the block. The constraint language can be formal (e.g., MATLAB or Java) or specified as part of the definition of the constraint. The par shows how the constraint properties are connected and bind to the value properties of the blocks. This diagram is a variant of the ibd that restraints the elements of the diagram to be constraint blocks, their parameters (bound to other parameters and properties of the blocks where they are used) and the block properties that they bind to.

The trade off studies are common in the field of Systems Engineering and are used to compare alternatives based on given criteria. A property that needs to be evaluated in a trade study is represented by a measure of effectiveness (moe) and a specialization of constraint block, defined as objective function (cost or utility function), is used to define how the solutions will be evaluated.

## Use Case Diagram (uc)

The use case diagram used in SysML is analogous to UML and is used to describe the high level functionality of a system in terms of how its users (called actors) use that system to achieve their goals. The uc is used to describe the relationships between the system under consideration (the subject), its actors, and use cases (the functionalities required by actors described graphically and textually). The detailed scenarios are usually described by means of other behavioural representations such as state machines (when the interaction between the actors and the subject is asynchronous), activities (when the scenario includes considerable control logic, input/output flows, and/or algorithms that transform data) or interactions (when a scenario is largely message based). The uc frame can represents a package, a model or a model library. This diagram is dedicated to model the system's behaviour in terms of functional requirements.

The actors may represent humans, organizations, or other external systems and are depicted either as a stick figure with the actor's name underneath the icon (Figure 6) or as a rectangle with the keyword "actor." The use cases describe the functionalities that a system must provide in order to accomplish the actors' goals. They are represented by ovals with the use case name inside. The system under consideration is represented by a block. A use case can represents several scenarios that correspond to the different paths the actors can follow through the use case under different circumstances.

Besides the graphical representation, the use case should be accompanied by a text based description captured by comments or SysML requirements. The textual description typically includes the preconditions, the postconditions, the primary flow, and the alternate/exception flows of the use case.

## Activity Diagram (act)

The SysML flow based behavioural modelling is accomplished through activities. The act is similar to the traditional EFFBD (SysML alignment with proven SE techniques), with additional features, and models the system behaviour in terms of transformation of inputs into outputs (I/O) through a controlled sequence of actions. The activities enhance the EFFBD providing capabilities to establish relationships with the structural elements of the system (e.g., blocks) and to model continuous flow behaviours. The activities are based on token flow semantics related to Petri nets. The activities may have multiple inputs and outputs called parameters. These can have types such as value types or blocks (e.g., a structural entity flowing through an assembly line), a direction (in or out, or both), and multiplicity that indicates how many tokens can be consumed or produced by each execution of the activity (the absence indicates multiplicity one).

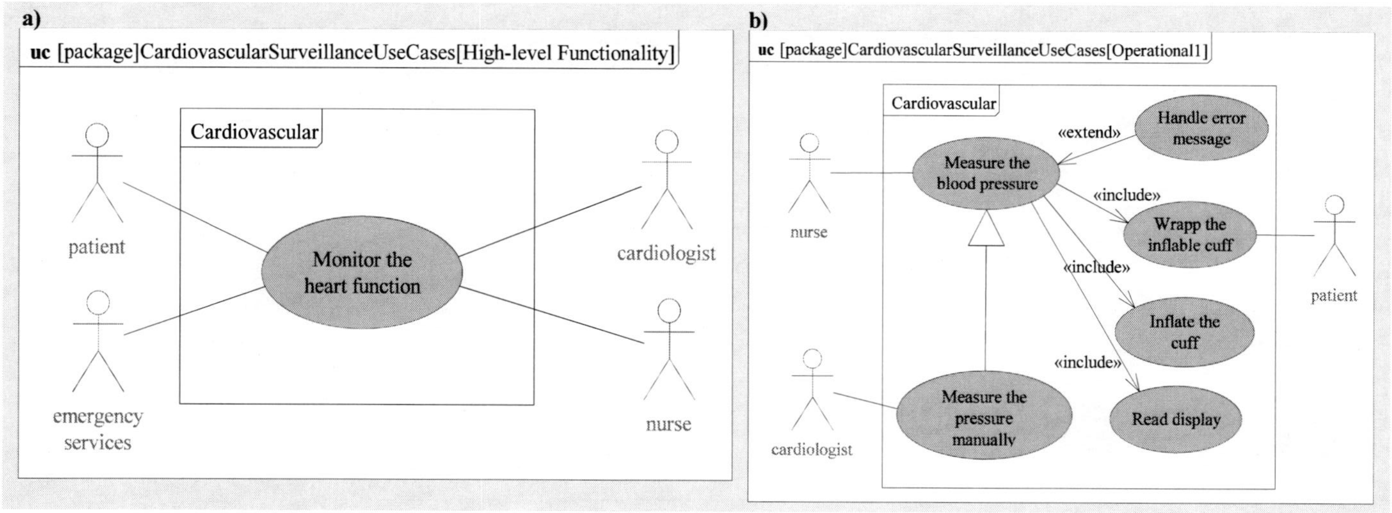


Figure 6. Use case modelling: a) uc for a high level functionality (use case 'Monitor the heart function') of the Cardiovascular Surveillance System and the actors that participate in it, b) uc for the operational use case 'Measure the blood pressure' that includes three other use cases, is extended by the use case 'Handle error message,' and has a specialized use case 'Measure the blood pressure manually.'

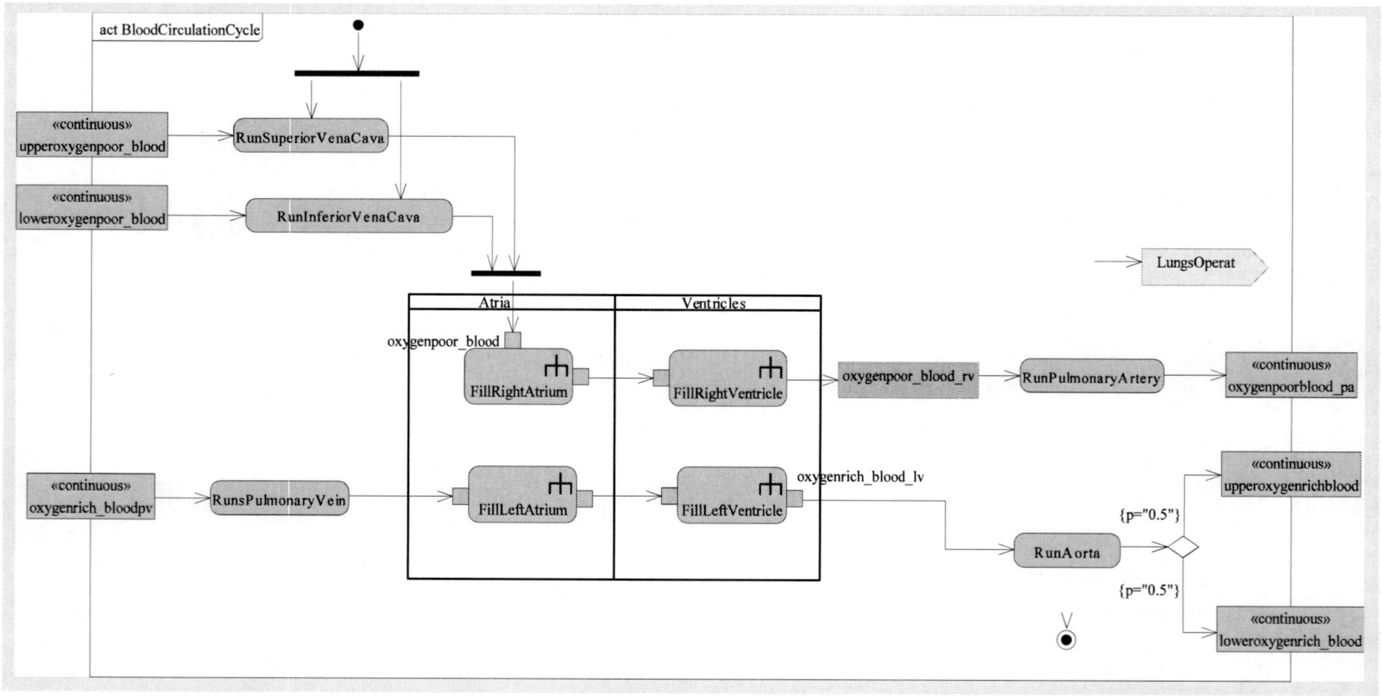


Figure 7. Simplified act for the heart circulation cycle depicting object and control flows, initial, activity final, fork, join and decision nodes, an object node (blue), a send signal action (yellow), and swimlanes to assign responsibility to the main heart parts (atria and ventricles).

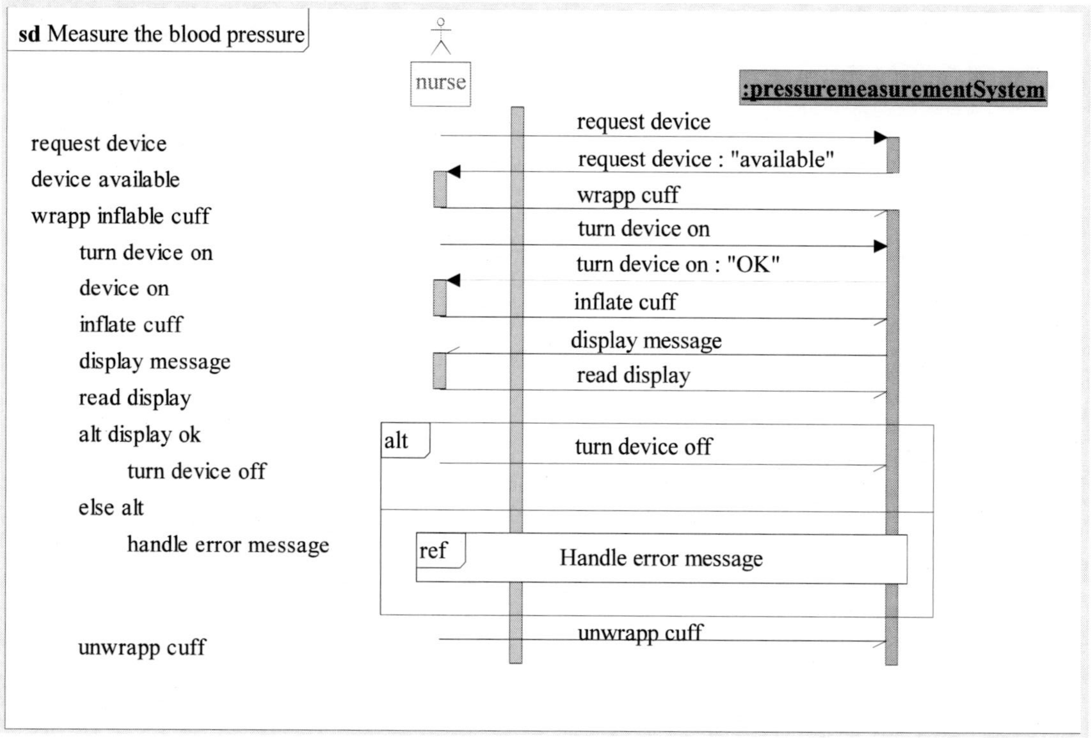


Figure 8. Simplified sd for the use case ‘Measure the blood pressure’ including an weak sequencing of occurrences, two lifelines, activations, a series of synchronous, asynchronous and reply messages, an alt operand of a combined fragment for the alternative courses of action derived from the displayed message at the measurement device, and an interaction use (frame ref) that specifies an interaction describe on other sd (Note: the Artisan Studio provides the description steps of the interaction on the left side of the diagram).

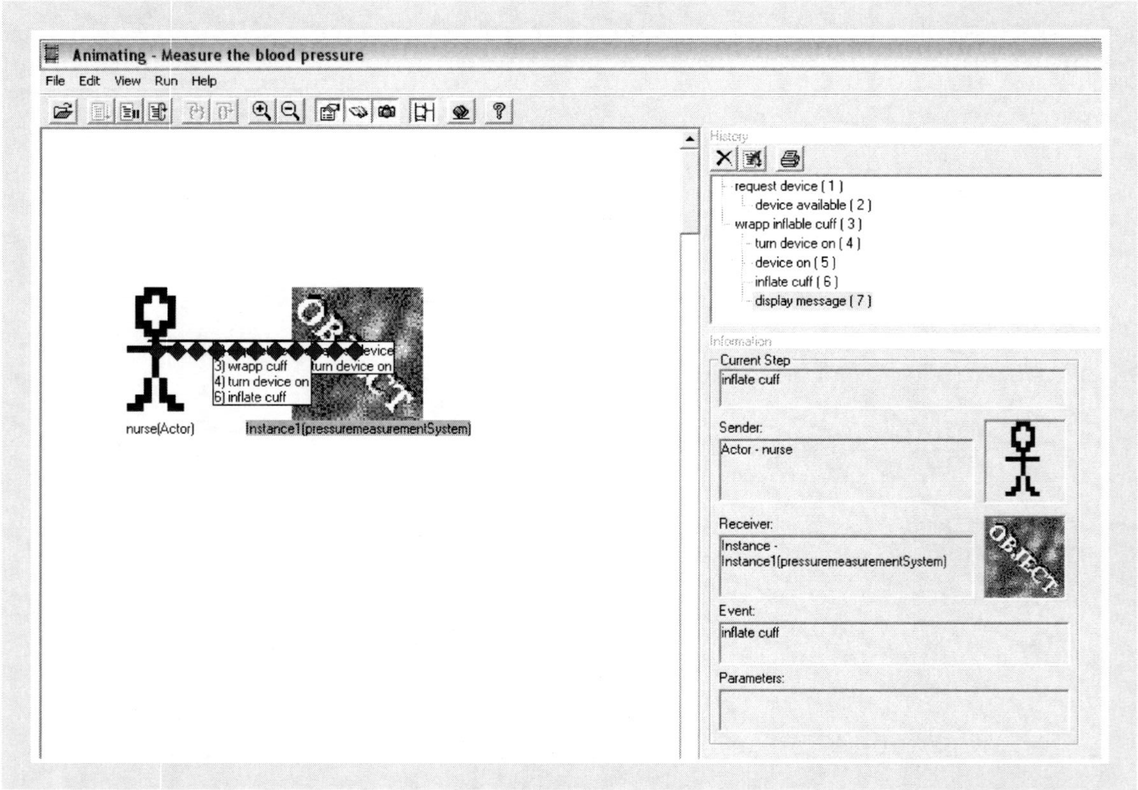


Figure 9. Animation snapshot for a sequence diagram.

The actions are the building blocks of activities and specify how they perform. The actions are the atomic level of activity behaviour and they can accept inputs and produce outputs (designated as tokens) on their pins. Flows can be discrete or continuous (the time between tokens is approximately zero). The actions have input and output pins (small rectangles acting like buffers) that accept tokens (units of flow) that may represent units of matter, information, energy and act as values of inputs, outputs, and control that flow from one action to another. The pins and activity parameter nodes are connected through object flows that enable the passage/route of tokens between actions and from/to parameter nodes. Figure 7 displays a simplified activity diagram for the heart circulation cycle.

The relationship between the behaviour of a system, expressed with activities, and the structure of a system, expressed using blocks, may be established through the utilization of, for example, activity partition or swimlane. This mechanism groups activity nodes, and in particular call actions, assigning responsibility for execution of those nodes. The swimlanes highlight the potential need for interfaces. The partitions are represented by rectangular symbols, aligned horizontally or vertically, encompassing the associated activity nodes. The rectangle has a header with the name of the model element (e.g., block, part, reference) represented by the partition (Figure 7). Other approach to connect behaviour and structure is to specify an activity (as main behaviour or method) in a block. An instance of the owning block executes the activity.

The functionality of an EFFBD (still largely used by systems engineers) can be represented as a constrained use of an activity diagram using the keyword "effbd" in the diagram header.

## Sequence Diagram (sd)

The behaviour of a system can also be expressed as a sequence of message exchanges that reflect the interaction between structural elements of a block. This kind of modeling is particularly useful to represent service-oriented behaviour, when a part of a system requests services of another part. The sd frame represents an interaction. This interaction corresponds to an ordered series of events between the structural elements of a block. The main structural feature of an interaction is the lifeline. This feature represents the lifetime of a member of the owning block that can be either a part property or a reference property. A part can be typed as an actor that can participate in interactions.

The lifeline is graphically represented by a rectangle (the head) with a dashed line descending from its base (the tail) ('nurse' and 'pressure measurement System' in Figure 8). When the interaction executes in an instance of its block, each lifeline represents an instance of the part of that block. So, when the part has multiplicity greater than 1, it can be used a selector expression (shown in square brackets after the name of the part) to identify one particular instance.

The time of an interaction progresses vertically down the diagram and the events on a lifetime are also ordered. Nevertheless, as is the case of distributed systems, a message may be overtaken by a subsequent message sent from the same lifeline and this can be represented by a downwards slanting arrow between two lifelines. The time can also be explicitly represented on sd by using time observations and duration observations and time and duration constraints involving the values of those observations.

The Artisan Studio tool provides a module, the Object Animator, which enables the animation of object interactions in a sequence diagram. This animation is very useful to explain the interactions to the end users. The Figure 9 shows a snapshot of the animation of the sd depicted in Figure 8.

## State Machine Diagram (stm)

The state machines describe behaviour in terms of the response of blocks to internal and external events. They support event based behaviour (generally asynchronous) that represents the lifecycle of a block. The behaviour is modelled as a set of states and the transitions between them. The diagrams are also known by state charts or state diagrams. As other SysML constructs, the sate machines can be composed hierarchically enabling complex representations. The state machines are typically owned by blocks and executed within the context of an instance of that block. The events experienced by the block may cause state transitions. Each state machine includes at least one region being each region defined in terms of states, pseudostates (termed vertices), and transitions between those vertices.

A state is a significant condition (usually expressed in terms of an implicit state variable) in the life of a block. The state specifies the effect of entering and leaving that condition and what the block does while it is in that condition. A transition describes a valid state change within a state machine and under what circumstance that change will happen. A transition has one or more triggers, a guard, and an effect. A trigger is associated to an event and identifies the possible stimulus that causes a transition to occur. The events may be of type: i) signal events (a new asynchronous message has arrived; the corresponding text is the name of the signal followed by an optional list of attribute assignments in parentheses), ii) time events (either a given time interval has passed since the current state was entered or a given instant of time has been reached; the corresponding text is “after” or “at” followed by the time in parentheses), iii) change events (some condition has been satisfied; the corresponding text is the term “when” followed by the condition that has to be met in parentheses), and iv) call events (an operation on the owning block has been requested; the corresponding text is the name of the operation followed by an optional list of attribute assignments in parentheses). A transition may also be triggered by a completion event when the current active state has completed.

A state machine may be reused to specify the behaviour of a submachine state (a state symbol showing the name of the state along with the name of the referenced state machine, separated by a colon). Interactions with the reused state machine occur via transitions to and from the boundary of the submachine state through entry point and exit point pseudostates.

The Figure 10 depicts a simplified state machine diagram for the block ‘pressure measurement system’.

## Requirement Diagram (req)

The requirements are text-based descriptions that can be related with a given capability or condition (that must be satisfied), with a function that a system should perform, or a performance condition that a system should achieve. The requirements typically reflect the needs of multiple stakeholders. These requirements can then be related to other requirements or to other model elements. They are cross cutting constructs since they support concerns that cut across the different system views. The “requirement” is a stereotype of the UML element class.

SysML provides a bridge between the standard requirements management tools and the system model keeping the requirements synchronized with the model. The requirements can be imported from the requirements management tool (e.g., Telelogic DOORS®), or text specification (e.g., Word®), or they can be created directly in the SysML tool. Systems engineers need to ensure that the requirements are consistent and feasible, have been validated by the stakeholders, and have been verified to ensure the adequacy of the system design. The use cases have been used to capture system requirements but they are particularly adequate to model functional requirements, being ineffective in the modelling of physical, availability, and other non-functional requirements. The SysML text-based requirements accommodate a considerable range of requirements' types that are mandatory in the field of SE.

The requirements are organized in hierarchies that describe a collection of requirements contained in a specification (containment hierarchy) organized in a tree structure. The req frame can represents a package or a requirement. The req is typically used to describe hierarchies of requirements (Figure 11) or to illustrate the relationships between a requirement and other model elements. Besides the req the requirements construct can be shown on bdd and uc. The relationship between requirements and other model elements can be depicted in bdd, ibd, and other diagrams through compartment and callout notations. The basic properties of a "requirement" model element are strings and are a unique identifier (id) and a descriptive text (text) (Figure 11). Additional properties such as verification status, requirements category, criticality, and risk can also be added. The categories of requirements should be adapted for the particular application and they can be created defining additional subclasses of the requirement stereotype. For example, a functional requirement may be constrained so that it can only be satisfied by an operation or behavioural element, a performance requirement can only be satisfied by a value property, a physical requirement may only be satisfied by a structural element, etc. SysML relationships to relate requirements include: "satisfy", "verify", "refine", "trace", "copy", "deriveReqt", and "containment". The containment relationship (a crosshairs symbol) is used to describe the hierarchical partition of a requirement (Figure 11). This partitioning (of complex into simpler) is fundamental to establish traceability and show how individual requirements are the basis for further derivation. The containment hierarchies can be used to create multiple levels of nested requirements. The requirements containment hierarchy is depicted in a compact way in the browser/explorer view of any modelling tool (top right corner of Figure 11).

The requirements relationships can be depicted through direct notation (when the requirement and the other model element it relates are in the same diagram; depicted as a dashed arrow with the name displayed as a "keyword" and pointing to the independent or supplier element), compartment notation (the element compartment displays the relationship and direction, e.g. satisfiedBy, and the model element name, e.g. pressuremeasurementSystem) (Figure 12), or callout notation (a comment symbol is attached to the model element displaying the same information on the compartment). It is only used one of these notations. SysML elements rationale (a comment notation with the keyword "rationale") and problem (a comment notation with the keyword "problem") can be added throughout the model to capture explanation for certain decisions or to emphasize particular problems. The text can make reference to external documents or other parts of the model. These elements can be attached to a requirement or to a requirement relationship (Figure 12).

The diagrams are not adequate to display a large number of requirements. Consequently, the majority of tools provide support to represent the requirements and their relationships through requirements tables and/or traceability matrices (Figure 13). The columns of the tables usually display the id and text of the requirement as well as the relationships. The matrices display the model elements in the rows and columns and the corresponding relationships in the cells. The matrices help to identify the origins, destinations, and links between requirements and the developed system's models.

## Other Cross-Cutting Constructs (Allocations)

The cross-cutting constructs support concerns that cut across the different views. Beside the requirements and parametrics, the allocations are a general mechanism that provides cross-cutting, supporting the interconnection of model elements from different model areas. Allocation is a term used by systems engineers to describe the organized cross-association (mapping) of elements within the different hierarchies or structures of the system's model. Allocations are typically used in early stages of the design process preceding more detailed specifications and implementations. The allocation relationship is used to map one model element to another. It provides an effective way to establish cross relationships and ensure that the various parts of the model are properly integrated. It can be established between any two model elements.

The notation used to represent allocation relationships can be graphical or tabular. The graphical notation includes the direct notation (when both model elements are at the same diagram they can be related through a dashed arrow with the keyword "allocate" and with the arrow pointing to the 'allocated to' end), the compartment notation (can be used for model elements that have compartments, such as blocks and parts, and identifies the element at the opposite end of the allocation relationship, depicting the keyword allocatedTo or allocatedFrom), and the callout notation (represented as an attached note symbol that specifies the name and type of the model element at the other end of the allocation relationship; the end is identified by the keyword allocatedTo or allocatedFrom). The tabular format is known as allocation table or matrix (Figure 14) and allows a compact representation of multiple allocation relationships.

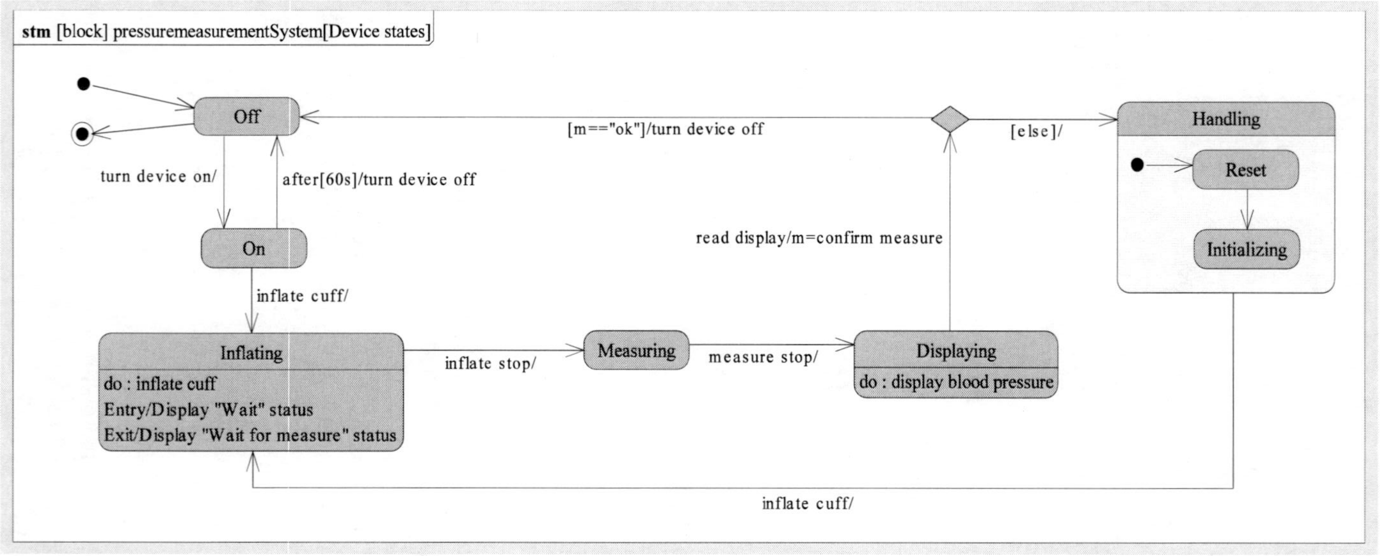


Figure 10. Simplified stm for the block ‘pressure measurement system’ including an initial pseudostate and a final state, some states with entry and exit behaviours and do activities, several triggers with signal events and one with a time event, guards, a choice pseudostate and one hierarchical state.

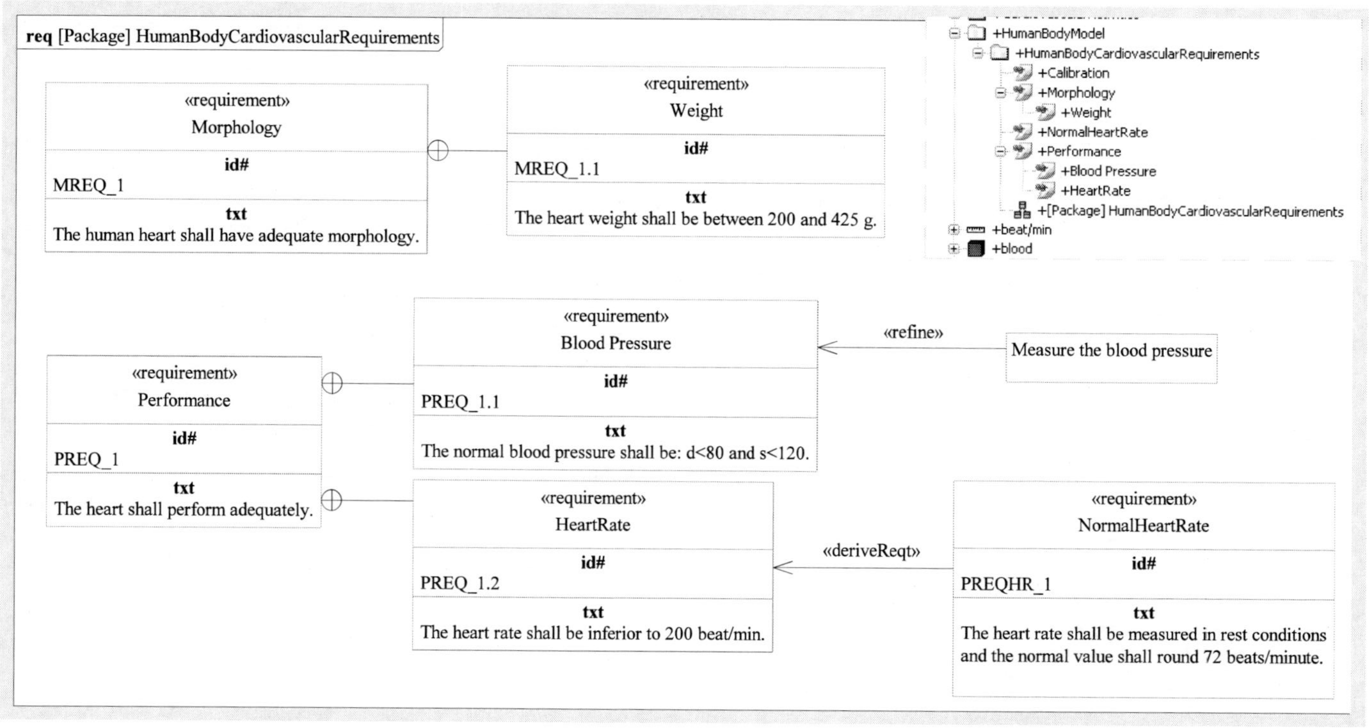


Figure 11. Simplified req for the human body cardiovascular requirements depicting the id and descriptive text for all the requirements, three containment relationships describing hierarchical partitioning, a derive relationship describing a more detailed requirement, and a refine relationship that clarifies the 'Blood Pressure' requirement through the sd 'Measure the blood pressure'; the top right corner depicts the requirements containment hierarchy presented by the explorer pane of the Artisan Studio tool.

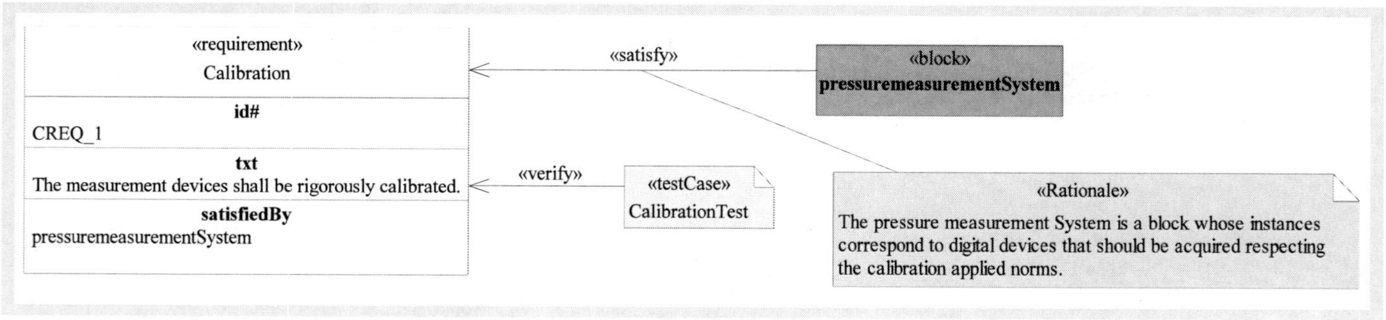


Figure 12. Example of a direct notation depicting a satisfy relationship and a compartment notation for the same relationship; the verify relationship is used to prove the calibration is satisfied by means of a Calibration Test (represents a sd); the satisfy relationship has a rationale that intends to capture the reason for the calibration needs.

**[Package] HumanBodyCardiovascularRequirements [Table1]**

| Name | Id# | Txt | Rationale | Satisfied By |
|---|---|---|---|---|
| Calibration | CREQ_1 | The measurement devices shall be rigorously calibrated. | | «block» pressuremeasurement: |
| Morphology | MREQ_1 | The human heart shall have adequate morphology. | | |
| Weight | MREQ_1.1 | The heart weight shall be between 200 and 425 g. | | |
| NormalHeartRate | PREQHR_1 | The heart rate shall be measured in rest conditions and the normal value shall round 72 beats/minute. | | |
| Performance | PREQ_1 | The heart shall perform adequately. | | |
| Blood Pressure | PREQ_1.1 | The normal blood pressure shall be: d<80 and s<120. | | |
| HeartRate | PREQ_1.2 | The heart rate shall be inferior to 200 beat/min. | | |

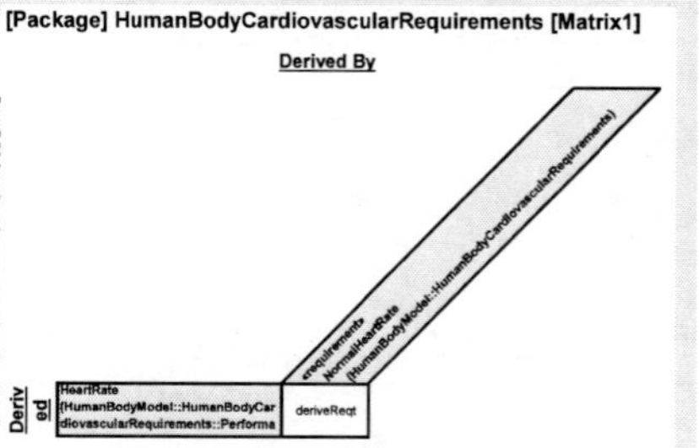


Figure 13. Example of a requirements table and a simple traceability matrix (with one cell).

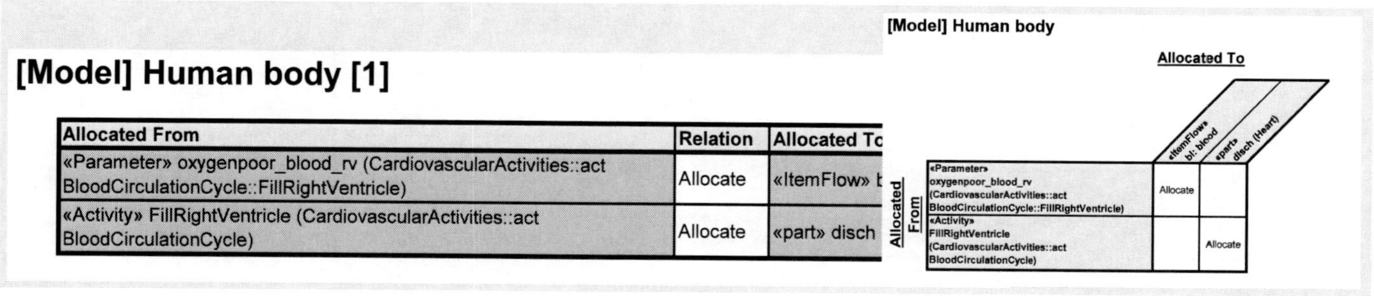


Figure 14. Example of an allocation table and an allocation matrix.

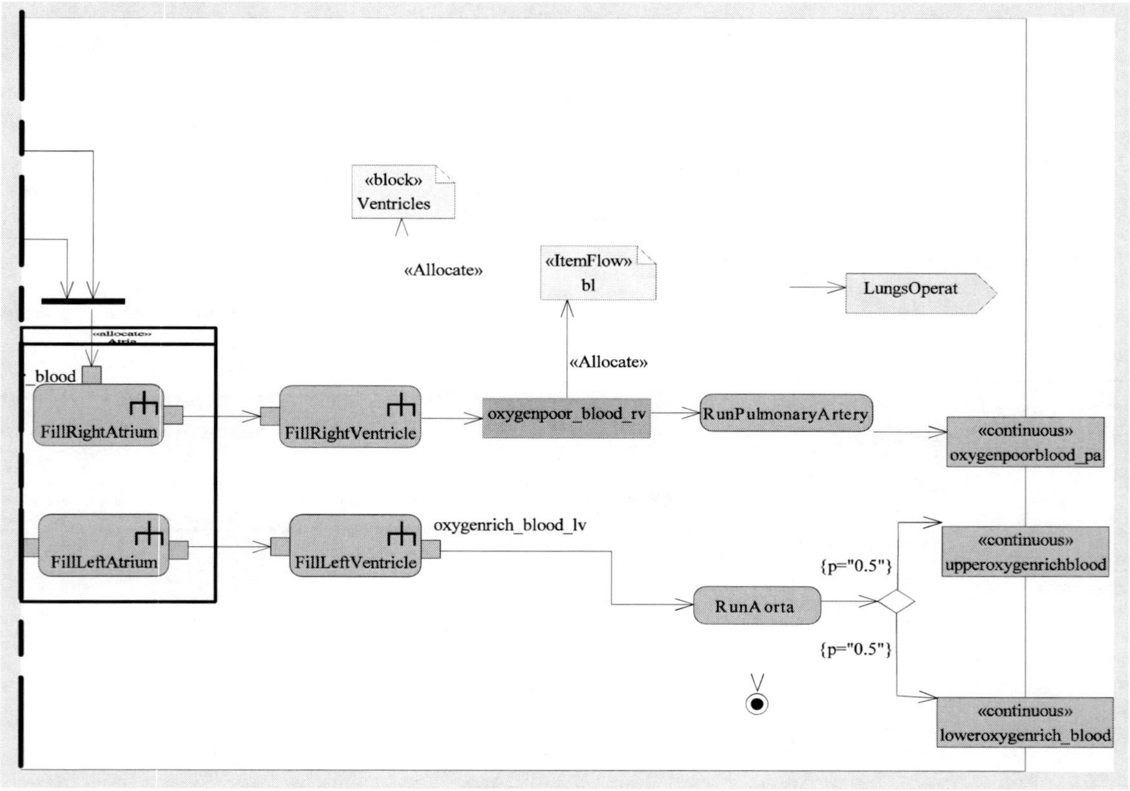


Figure 15. Region of an act depicting a functional allocation and a flow allocation.

The allocations can be of different types including the behavioural/ functional (e.g., allocating a function to a component), the structural (e.g., allocating logical elements to physical devices or software modules to hardware components), the requirements, the properties, and the flow ones.

The requirements allocation is typically used to map source requirements to other derived requirements, or to map requirements to other model elements that satisfy the requirement (Figures 11 and 12). The properties allocation is used to allocate performance or physical properties to different elements of the system. This kind of allocation is usually done by means of parametric constraints from value properties to parameters (Figure 5). The flow allocation is used to allocate flows between activity diagrams and internal block diagrams (Figure 15 depicts a flow allocation from the object flow oxygenpoor_blood_rv to the item flow blood depicted in the ibd of Figure 4).

The behavioural/functional allocation is a mechanism used to separate the models of form (or structure) from the models of behaviour so that the design options can be selected by considering different structural alternatives that provide the desired emergent behaviour (to optimize designs based on trade studies among the different alternatives looking at the relationships between behaviour and structure for each alternative). The structural allocation is used to allocate elements between structural models. The logical physical allocation (e.g., a logical block hierarchy is mapped to another physical block hierarchy at a more concrete level of abstraction) and the software hardware allocation (e.g., the software artifacts are deployed to hardware platforms or processing nodes) are typical examples of this type.

The allocations can be evaluated according to their completeness, consistency, and overall balance. For example, in functional allocation, the allocation of a set of activities is complete when each activity has an allocation relationship to a block in the model. The consistency can be checked when the action nodes are depicted in a valid act, or when the object flows are allocated to appropriate connectors in an ibd. The automated model checking provided by the majority of the tools is expected to help in this verification avoiding circular, redundant, and inappropriate allocations. The balance is more subjective and involves the evaluation of the different levels of abstraction as well as the degree of allocation. For example, if a single block has a considerable number of allocated activities the modeller might consider the decomposition of that overallocated block that integrates too much functionality.

As UML, SysML is not attached to any methodoly. The language supports data interoperability through the XML Metadata Interchange (XMI) standard and the ISO 10303 AP-233 standard (or STEP, the Standard for the Exchange of Product model data) for exchanging and sharing information during the whole system development lifecycle and across different domain engineering disciplines), and also provides mechanisms for customization and domain-specific modeling such as libraries and profiles.

## CONCLUSION

The MBSE field claims for convergent standards and domain-specific modelling languages able to reflect a new way of thinking and working. In practice, this way of working is formalized through the use of models by collaborative world teams joining their efforts from different places around the globe. Obviously, they must "speak" the same language and must work on the same "matter." The "matter" is the *System Model* and the communications mechanisms must be supported by standard, flexible, and friendly graphical modeling languages such as SysML (Ramos et al., 2012).

Real case studies are fundamental to understand the modern challenges and to prove the value of the language. It is our opinion that this empirical work will have as target domain the complex super systems that aim to deliver world sustainability and respond to societal needs. The main challenge will be to ensure that the *System Model* reflects the stakeholders' ideas and positions acting as a shared working platform, and the resulting System satisfies their expectations. This tasks is becoming more complex and challenging once the stakeholders are demanding increased functionality, reliability, lower prices, shorter life cycles, and environamentally and socially acceptable solutions with value to the users, i.e., sustainable, scalable, safe, smart, stable, simple, and secure systems (BKCASE Editorial Board, 2016).

The emergence of the MBSE discipline is today well-visible in the new dedicated conferences that flourished during the last years (e.g., International Conference on Systems Engineering and Modelling sponsored by the IEEE, the ACM/IEEE International Conference on Model Driven Engineering Languages and Systems). The topics of interest in these conferences include, for example, 'Standards for SE models,' 'UML and SysML: scope, expectations, and observations', 'Evolution of modelling languages,' 'Object-Process Methodology,' 'Methodologies, processes and modelling languages,' 'Comparative studies of conceptual modelling approaches and languages,' and 'New types of SE paradigms: biological, nano'.

The authors believe that SysML can be, effectively, a graphical tool to support the work of multi-disciplinary and globally disperse working teams presenting a rich and comprehensive standard notation already supported by several commercial tools.

## REFERENCES


Bjorkman, E., Sarkani, S. and Mazzuchi, T. (2013). Using Model-Based Systems Engineering as a Framework for Improving Test and Evaluation Activities. *Systems Engineering, 16*(3), 346-362.

BKCASE Editorial Board (2016). The Guide to the Systems Engineering Body of Knowledge (SEBoK), v. 1.6. R.D. Adcock (EIC). Hoboken, NJ: The Trustees of the Stevens Institute of Technology. Accessed May 2016. www.sebokwiki.org.

Buede, D. (2009). *The Engineering Design of Systems-Models and Methods,* 2nd edition. New Jersey: John Wiley & Sons, Inc.

David, P. Idasiak, V. and Kratz, F. (2010). Reliability study of complex physical systems using SysML. *Reliability Engineering and System Safety, 95(4)*, 431-450.

Dori, D. (2002). *Object-Process Methodology: A Holistic Systems Paradigm*. New York: Springer.

Durugbo, C. (2013). Integrated Product-Service Analysis UsingSysML Requirement Diagrams. *Systems Engineering, 16(1)*, 111-123.

Friedenthal, S., Moore, A. and Steiner, R. (2008). *A Practical Guide to SysML, The Systems Modeling Language*. Burlington: Morgan Kaufmann/OMG Press, Elsevier Inc.

Friedenthal, S. (2009). SysML: Lessons from Early Applications and Future Directions. *INSIGHT-INCOSE Journal, 12*(4), 10-12.

Gomes, R., Hoyos-Rivera, G., Willrich, R., Lima, C. and Courtiat, J-P. (2011). A Loosely Coupled Integration Environment for Collaborative Applications, *IEEE Transactions on Systems, Man and Cybernetics, Part A – Systems and Humans, 41*, 905-916.

Grobshtein, Y. and Dori, D. (2009). Creating SysML Views from an OPM Model. *Proceedings of the 2nd International Conference on Model-Based Systems Engineering MBSE'09*, Herzelya and Haifa - Israel, 36-45.

Hughes, R. (1999). Models and Representations. *Philosophy of Science, 64*, 325-336.

Karcanias, N. (2004). Modelling and simulation in technological and emerging fields: emerging challenges. *White Paper*, Workshop: SIM-SERV Modelling and Simulation Challenges, Working Group Roadmap for Continuous and Hybrid Simulation, University of Patras, Greece. Available at http://www.sim-serv.com/wg_doc/WG7_General_roadmap.pdf.

Law, A. (2014). *Simulation Modeling and Analysis*. McGraw-Hill Education.

Ludewig, J. (2003). Models in software engineering – an introduction. *Software and Systems Modelling*, 2(*1*), 5-14.

Oliver, D., Andary, J. and Frisch, H. (2009). Model-Based Systems Engineering, in *Handbook of Systems Engineering and Management*, 2nd edition, Eds. Sage. A. and Rouse W., 1361-1399, New Jersey: John Wiley & Sons, Inc.

OMG SysML (2016). OMG Systems Modeling Language: The Official OMG SysML site. Accessed in may 2016 [http://www.omgsysml.org/].

Ramos, A., Ferreira, J. and Barceló, J. (2012). Model-Based Systems Engineering: An Emerging Approach for Modern Systems. *IEEE Transactions on Systems, Man, and Cybernetics, Part C – Applications and Reviews, 42*(1), 101-111.

Ramos, A., Ferreira, J. and Barceló, J. (2013). LITHE: An Agile Methodology for Human-Centric Model-Based Systems Engineering. *IEEE Transactions on Systems, Man, and Cybernetics: Systems, 43*(3), 504-521.

Rhodes, D. (2008). Addressing Systems Engineering Challenges Through Collaborative Research. *SEARI - Systems Engineering Advancement Research Initiative*. Massachusetts Institute of Technology.

Ross, D. (1977). Structured Analysis (SA): A Language for Communicating Ideas. *IEEE Transactions on Software Engineering*, *3(1)*, 16-34.

Rumbaugh, J., Jacobson, I. and Booch, G., (1999). *The Unified Modeling Language Reference Manual*. Massachusetts: Addison Wesley Longman, Inc.

Sánchez, D., Cavero, J. and Marcos, E. (2009). The concepts of model in information systems engineering: a proposal for an ontology of models. *The Knowledge Engineering Review*, *24(1)*, 5-21.

Senge, P. (1990). *The fifth discipline – The Art & Practice of the Learning Organization*. New York: Doubleday Business.

Sommerville, I. (2010). *Software Engineering*, 9th edition. London: Pearson Education Limited.

Sussman, J. (2000). *Introduction to Transportation Systems*. Massachusetts: Artech House, Inc.

Tien, J. (2008). On Integration and Adaptation In Complex Service Systems. *Journal of Systems Science and Systems Engineering*, *17*(4), 385-415.

Van Daalen, C., Thissen, W. and Verbraeck, A. (2009). Methods for the Modeling and Analysis of Alternatives, in *Handbook of Systems Engineering and Management*, Eds. Sage A. and Rouse W., 1037-1076, New York: Wiley-Interscience.

Whitten, J., Bentley, L. and Dittman, K. (2004). *Systems Analysis and Design Methods*. New York: McGraw-Hill/Irwin.

Wilkiens, T. (2007). *Systems Engineering with SysML/UML – Modeling, Analysis, Design*. Burlington: Morgan Kaufmann/OMG Press.

In: Systems Engineering
Editor: Marlene Hopkins

ISBN: 978-1-63485-752-9


*Chapter 6*

# COMPLEX SYSTEM ANALYSIS FOR ENGINEERING OF SYSTEMIC FAILURES

***Polinpapilinho F. Katina*[1,*] *and Behnido Y. Calida*[2]**
[1]Department of Engineering Management and Systems Engineering,
National Centers for System of Systems Engineering,
Old Dominion University, Norfolk, VA, US
[2]Center for Data Reduction and Analysis Support,
Virginia Tech Transportation Institute, Virginia Tech, Blacksburg, VA, US

## ABSTRACT

At the very fundamental level, the utility of engineering approaches is to offer greater understanding of a given phenomenon so as to enable problem solving in real world situations. However, all engineering approaches are not of equivalent value – there are different benefits and limitations associated with each methodological approach for engineering. This chapter examines the nature of differences. First, a high level classification of methodology approaches is provided. The aim is this section to 'bin' different engineering methods based on how one obtains and communicates knowledge (i.e., underlying philosophy). Second, the type of failures that must be addressed by systems-based approaches are examined. This section seeks to provide a context-independent set of system failure paradigms that can be used to deal with different failures in complex systems. A context-independent

[*] Corresponding author: Polinpapilinho F. Katina. Email: pkatina@odu.edu.

conceptualization of systemic failures would include formulating an acceptable level of abstraction of the system. This ensures that underlying worldviews are made explicit and appropriately represented to reveal system pathologies. To help address this, a systems theory-based approach using system pathologies and the M-Path Method™ is introduced. This provides a framework through which perceived systems of interests can be analyzed to identify systemic failures. The chapter concludes with practical insights for how to effectively deal with complex system failures – including recent tools.

**Keywords**: 21st century, hard systems thinking, M-Path Method, problem formulation, soft systems thinking, systemic failure, systems engineering, systems theory, systems theory-based methodologies, systems theory-based pathologies

## INTRODUCTION

In science, the combination of procedures for gaining knowledge about phenomena and structured processes involved in intervening and perhaps changing a given system of interest is often referred to as a methodology (Burrell and Morgan, 1979; Checkland, 1993; Churchman, 1968; 1971). Arguably, the need for methodological approaches is related to need to solve real-world problems. Interestingly, there is no shortage of methodological approaches supporting different problem domains such as Complex System Engineering, Enterprise Systems Engineering, Complex Adaptive Systems Engineering, and System of Systems Engineering (Sousa-Poza, 2015). These methodological approaches are often developed when present methodologies are found to be inadequate and/or faced with emergence of a new problem domain. For instance, take the case for System of Systems Engineering. It emerged due to the "the need to better understand how existing, legacy, sociotechnical systems and new components can be brought together into a structure that leverages available resources to generate new capabilities" (Sousa-Poza, 2015, p. 162). A complementary view of the need for different methodological approaches is offered through the lens of time. Literature indicates that modern times are characteristically tumultuous (Martin, 2006; Tainter, 1988) aligning with the notions of 'messes' (Ackoff, 1974) as well as 'wicked problems' (Rittel and Webber, 1973). This characterization, elaborated in Table 1, is manifested in elements of ambiguity, complexity, emergency, interdependencies, and the uncertainties of the last three centuries (Keating et al. 2014).

These characterizations highlight an ongoing need for robust and yet unique methodologies (Conrad and Gheorghe, 2011). However, the calls for methodological robustness and uniqueness suggests that there is also a need to match methodological approaches to problem domains. This chapter is especially designed to address the matching of methodological approach to problem domain. First, a high level classification of methodologies associated with engineering are provided. The aim of this section is to 'bin' methodologies based on their underlying philosophy. Ultimately, an effective methodological approach helps in gaining knowledge about phenomena and provides an interim structure to intervene in a given system. An important element of this aspect of methodology is the necessity to deal with systemic failures. A second element of this chapter provides a classification of systems failures. The nature of failure is discussed to offer a perspective of the relationship between methodological approach and failures, methods, and tools. The chapter concludes with practical insights for how to effectively deal with systemic failures.

**Table 1. Characteristics of modern landscape**

| Characterization of modern times | Brief description |
|---|---|
| *Ambiguity* | There is increasing lack of clarity and situational understanding of complex systems |
| *Complexity* | Large numbers of richly and dynamically interacting complex systems and subsystems with behavior difficult to predict |
| *Emergence* | Inability to deduce behavior, structure, or performance of complex systems as a function of their constituent subsystems and elements |
| *Interdependence* | Mutual influence among different complex systems through which the state of a system influences, and is influenced by, the state of other interconnected systems |
| *Uncertainty* | Uncertainty appears to be the norm in which there is always incomplete knowledge about situations and therefore casting doubt for decision/action consequences |

## SYSTEMS-BASED METHODOLOGIES

It is necessary to articulate the foundations of methodologies prior to engaging a discourse on methodologies themselves. First and foremost, the methodologies explored in this examination are taken from the Systems Thinking Paradigm. Contemporary researchers propose Systems Thinking as a science of systems that aims to understand the interaction between man and the environment from multiple perspectives (Skyttner, 2005). Although Systems Thinking does not have a single common and accepted definition, it suffices to say that it emerged as an alternative response to reductionism, which is closely aligned with the 'scientific method' (Katina, 2015a; Laszlo, 1996). Conceptualization of Systems Thinking, based in Systems Theory, gained momentum in early 1920s as doubts emerged about the continuing utility of the classical scientific approach of isolating constituent elements. The shift away from reductionism resulted in more interest being paid to notions of whole 'organizations' as opposed to parts (von Bertalanffy, 1972). Ludwig von Bertalanffy, a proponent of Systems Theory, suggested that "[s]ince the fundamental character of the living thing is its organization, the customary investigation of the single parts and processes cannot provide a complete explanation of the vital phenomena. This investigation gives us no information about the coordination or parts and processes" (von Bertalanffy, 1972, p. 410).

It was later noted by another proponent of Systems Theory that the traditional scientific approach, while relevant, creates "closed bubbles [of knowledge] in their own right. [Where] specialists in one field can communicate with one another if they share a specialty, but experience difficulty when their interests do not coincide. [However] the unfortunate consequence of such specialty barriers is that knowledge, instead of being pursued in depth and integrated in breadth, is pursued in depth in isolation" (Laszlo, 1996. p. 2). The same sentiments are echoed by Debora Hammond who remarks: "problems confronting humanity at this stage in our history (poverty, violence, crime, environmental degradation and nuclear weapons… terrorism) are systemic and cannot be understood or resolved in isolation. They result from…perception, rooted in the mechanistic worldview we inherited from the scientific revolution of the seventeenth century...we needed a more ecological or systemic world, based on an understanding of our fundamental interconnectedness and interdependence, with each other and with all of life" (Hammond, 2002, p. 430).

The last statement appears to echo a statement attributed to Albert Einstein: *we cannot solve our problem with the same thinking we used when we created them.* An examination of three different yet related aspects of systems theory, systems science, systems technology, and systems philosophy, helps to further distinguish the field of General Systems Theory[1]. Table 2 elaborates on these elements of Systems Theory. Ludwig von Bertalanffy posited that part of the solution to the reductionism problem generated by the Scientific Method was a need to 'discover laws at all levels of organization in biological systems' (von Bertalanffy, 1972) as a means to gain knowledge about entities and how they are related.

**Table 2. Aspects of Systems Theory**

| Aspects of Systems Theory | Description of the element of Systems Theory |
|---|---|
| *Systems Science* | This is the aspect of systems theory that deals with knowledge of the connected 'wholes' - complexity as opposed detailed and isolated systems (Katina, 2015b). von Bertalanffy (1972, p. 414) describes systems science as the "scientific exploration and theory of 'systems' in various sciences (e.g., physics, biology, psychology, social sciences), and general systems theory as the doctrine of principles applying to all (or defined subclasses of) systems." |
| *Systems Technology* | This is the aspect of systems theory that deals with "problems arising in modem technology and society, including both 'hardware" (control technology, automation, computerization, etc.) and 'software" (application of systems concepts and theory in social, ecological, economical, etc., problems)" (von Bertalanffy, 1972, p. 420). This aspect of systems theory is directly related to the need for a science of 'systems' in which emerging world issues such as pollution, economies, health, politics, and international conflicts had to be addressed from a holistic perspective (Katina, 2015b; Strijbos, 2010; von Bertalanffy, 1972; Warfield, 1976) |
| *Systems Philosophy* | An aspect of systems theory dealing with philosophical issues related to paradigm change within which systems theory supposedly operates (Katina, 2015b). Three elements epitomize this aspect of systems theory: 1) systems ontology, which deals with how an observer views reality. 2) Systems epistemology, which deals with how one obtains and communicates knowledge, and 3) nature of man in which man could be viewed as determinist or voluntarist |

[1] General Systems Theory was the original term proposed by Ludwig von Bertalanffy. However, it is common understanding that a general systems theory has yet to emerge. For more discussion on this topic see Adams (2012), Adams et al. (2014), Gaines (1977), Katina (2015b), and Monod (1974).

In summary, *Systems Science* is concerned with system laws, principles, and theorems that govern complex systems (Adams et al. 2014; Hester and Adams, 2014; Jackson, 2003; Katina, 2015a; Katina, 2015b; Katina, 2016; Whitney et al. 2015). *Systems Technology* is concerned with holistic approaches that could be undertaken to address current vexing issues from a perspective of systems theory. *Systems Philosophy* is concerned with the role of paradigm in addressing complex systems. It is a common understanding that the paradigm for systems theory is ontology-nominalism, epistemology-antipositivist with a concern for 'systems' as they are influenced by human values (Burrell and Morgan, 1979; Flood and Carson, 1993; Katina, 2015b; von Bertalanffy, 1972).

The preceding aspects of systems theory underscores the importance of worldview as well as corresponding approaches. These approaches must be undertaken to understand the natural world along with the governing laws and principles. Appreciation of the issues calling for systemic approaches is necessary if we are to develop methods and tools that can be used to intervene and bring about positive change in real-world of complex systems. Methodologically, there is no shortage of methods based on the articulated notions of systems theory. A methodology is taken as a set of "procedures for gaining knowledge about systems and structured processes involved in intervening in and changing systems" (Jackson, 1991, p. 134). It is important to recognize two issues in relation to methodology. First, there are two opposing extremes of methodological approaches, idiographic and nomothetic as suggested by Burrell and Morgan (1979). Second, there is a tendency to emphasize application of the idiographic view of a methodology in support subjectivity in research associated with complex systems (Flood and Carson, 1993; Katina et al. 2014a).

A further categorization of systems-based methodologies clarifies such approaches as hard and soft. A hard systems approach "assume[s] that problems are set in mechanical-unitary contexts. Hard methodologies take it as a given that it is relatively easy to establish clear objectives for the system in which the problem resides - so context must be unitary. They then try to represent that system in a quantitative model that simulates its performance under different operational conditions - something only possible if the system is simple and the context mechanical" (Jackson, 1991, p. 30).

However, since "it is often difficult to define precise objectives on which all stakeholders can agree" especially in complex systems (Jackson, 2003, p. 20), soft systems thinking approaches emerged to accommodate multiple and sometimes conflicting values, beliefs, and worldviews that are prevalent in complex systems. To support a needed change, "the solution was to make subjectivity central, working with a variety of world views during the methodological process" (Jackson, 2003. p. 22). Hence, a 'soft' systems approaches recognize that there can be a "vast numbers of relevant variables and the myriads of interactions make this [optimization] an impossible requirement. The solution... [is] to identify those key mechanisms or structures that govern the behavior of the elements or subsystems ...aspects that lie behind system viability and performance" (Jackson, 2003, p. 21). Table 3 provides a summary of 'hard' and 'soft' systems-based methodologies.

There are several areas in which systems theory-based methodologies can offer utility. These include aspects of learning, understanding, intervention, and ultimately bringing about positive change (Katina, 2015b; Kovacic et al. 2007). This notion might be referred to as governance (Calida, 2013). Certainly, governance embodies aspects that support "grappling with complexity [and] has to be a methodology for human learning" (Warfield, 1976, p. 2) and robust problem formulation grounded in systems theory (Katina, 2015b). However, taking into account the entire operational landscape of systems and the nature of current systems as described in Table 1, we suggest that great difficulty continues to exist in selection of an efficient means to achieve a known objective and a defined end. This is where the unity of systems-based methodologies is supposed to provide utility.

A critical aspects of systems theory-based methodologies, articulated in Table 3, is dealing with systemic failures. The nature of failures and systemic tools that can be used in dealing with failures is the subject of the remainder of this chapter.

## SYSTEMIC FAILURES

There is no shortage of literature supporting the notion that there is increasing occurrence of disastrous events (Ansoff, 1984; Cohen and Ahearn, 1986; Richardson, 1994; Weick, 1988). The problem is not limited to occurrence. Another extended assertion postulated by Rasmussen and Batstone (1989, p. ii) is that: "the frequency and magnitude of organizational failures and the subsequent impacts are increasing at an alarming rate." This notion has been supported by countless researchers as well as annual analysis of Swiss Reinsurance[2].

---

[2] More information can be found at http://www.swissre.com/.

**Table 3. A Summary of 'hard' and 'soft' systems-based methodologies**

| Classification | Methodology | Description | Primary proponents |
|---|---|---|---|
| *Hard Systems Thinking* | Systems Analysis | Intrinsically related to systems engineering and mechanistic in nature, this methodology is largely dependent on feedback loops and black boxes of cybernetic management to optimize socio-technical systems based on fixed parameters such as cost and benefits. A number of phases associated with this methodology are discussed elsewhere (Miser and Quade, 1988; Gibson et al., 2006). | Atthill (1975); Digby (1989); Gibson et al. (2007) |
| | Systems Engineering | The approach focusses on defining technical and business customer needs with the goal of producing quality products that meet user needs. A generic life-cycle model associated with this approach along with its stages are discussed elsewhere (Blanchard and Fabrycky, 2006; Componation and Collopy, 2012; Smartt and Ferreira, 2012) | INCOSE (2011); Blanchard and Fabrycky (2006) |
| | Operations Research | This approach is commonly associated with determining a maximum or minimum variable (e.g., profit, performance, yield, loss, risk) inventory, allocating, waiting-time, replacement, competitive, and combined processes, operations research was developed to deal with complex organizations that are under control of management (Churchman et al., 1957; Jackson, 2000). A generic model associated with this approach is discussed elsewhere (Churchman et al., 1957; Jackson, 2000). | Churchman. Ackoff and Arnoff (1957) |
| *Soft Systems Thinking* | Systems Dynamics | This methodology is concerned with limits of growth and understanding of the system structure using feedback loops as the main determinants of system behavior. Mathematical in nature, this approach involves four major variables: system boundary, network of feedback loops, variables of 'rates' or 'flows' and 'levels' or 'stocks,' and leverage points. | Forrester (1961); Sterman (2000) |
| | Organizational Cybernetics | This approach embodies the idea that organizations are black boxes characterized by complexity, self-regulation and probabilistic behaviors. Central to this approach is the Viable System Model, which is based on neurocybernetic model, consisting of five essential subsystems that are aligned with major viable organizational functions. VSM is a model rather than a methodology, it does not have a clear set of prescribed phases for deployment. Two general stages of system identification and system diagnosis are discussed elsewhere (Jackson, 2003; Katina, 2015b) | Beer (1979; 1981; 1985) |

| Classification | Methodology | Description | Primary proponents |
|---|---|---|---|
| *Soft Systems Thinking* | Strategic Assumption Surfacing and Testing | Methodological approach grounded on the premise that formulation of right solutions to the right problem requires uncovering critical assumptions underlying policy, plan, and strategy. This enables management to compare and contrast and gain new insights on their assumptions and 'wicked' problems. | Mitroff and Emshoff (1979); Mason and Mitroff (1981) |
| | Interactive Planning | Developed by Russell L. Ackoff, this methodology focuses on creating a desired future by designing desirable present conditions. It is made up two parts: *idealization* and *realization*. These parts are divisible into six interrelated phases. | Ackoff (1974; 1981a; 1981) |
| | Soft Systems Methodology | Attributed to Peter Checkland and his colleges at Lancaster University, this methodology emerged as a response to need for methods that can be used to intervene in 'ill-structured' problem situations where it is important to learn about systems while still focusing on 'goal-seeking' endeavors that answer 'what' should be done and 'how' it should be done (Jackson, 2003). Checkland's (1990) work suggested that understanding issues such as context, largely ignored in systems engineering provided a more rigorous attempt to tackle problematic situations coherently. | Checkland and Scholes (1990); Wilson (1984) |
| | Systems of Systems Engineering Methodology | This methodology is intended to provide a high-level analytical structure to explore complex system problems (Adams and Keating, 2011). In order to enhance our understanding of complex systems, SoSEM is taken as a "rigorous engineering analysis that invests heavily in the understanding and framing of the problem under study" (Adams and Keating, 2011, p. 113). | Adams and Keating (2009; 2011); Keating et al. (2004) |
| | Critical Systems Heuristics | Developed by Werner Ulrich, this methodology is concerned with 'unfairness in society' (Jackson, 2003). It promotes emancipatory systems thinking for planners and citizens alike. Synonymous with this methodology are three phases. | Ulrich (1983; 1987) |
| | Organizational Learning | This methodology is concerned with single-loop and double-loop learning where management of organization is able to contrast 'expected outcomes' with the 'obtained outcomes.' Contrasting these outcomes involves learning based on errors discovered during single-loop learning and provides the basis for modifying organizational norms, policies, and objectives (Fiol and Lyles, 1985). A key premise of this methodology is that learning and adapting new knowledge must be generated at the individual as well as at organizational level. | Argyris and Schön (1978; 1996) |

**Table 3. (Continued)**

| Classification | Methodology | Description | Primary proponents |
|---|---|---|---|
| *Soft Systems Thinking* | Sociotechnical Systems | Attributed to Eric Trist, Ken Bamforth and Fred Emery and their work at the Tavistock Institute in London, this methodology is concerned with a joint optimization of both social/soft including human and technical aspects of organizations (Pasmore, 1988). | Trist and Bamforth (1951); Cherns (1976) |
| | Total systems Intervention | This meta-methodology emerged out of the recognition of strengths of capabilities of individual systems approaches, the need for pluralism in systems thinning, and calls for emancipatory ideas in systems thinking – in reference to critical systems thinking (Jackson, 2003). It is based on the premise that contemporary systems-based methodologies are not complementary. This thinking aligns with a suggestion that a successful complex organizational intervention might require a 'combination' of any set of systems-based approaches Laszlo and Krippner (1998). | Flood and Jackson (1991); Flood (1995); Jackson (1991) |
| | Complex System Governance | An approach based on the design, execution, and evolution of nine metasystem functions that provide for the control, communication, coordination, and integration of a complex system. | Keating (2015); Keating and Bradley (2015); Keating and Katina (2016) |

Rather than revisiting the notion of failure from the traditional approach of probability and consequence (Katina et al. 2014b), present efforts are dedicated to the classification of systemic failures intrinsically related to the nature of operating landscape of complex systems. Needless to say, we suggest failures can be related to the following assertions:

1. Failue is a loss of ability to satisfactorily achieve intended function, that encompass fracture and/deterioration
2. Failure is a state of condition of not meeting or unacceptably deviating from specified or implied performance requirements
3. Failure include events that might render a system no longer capable of operation
4. Failures lessens viability (i.e., maintenance of separate existence) of a system of interest
5. Failure includes subjectivity of the observer and therefore can be a matter of interpretation and perception

This view of failure can be categorized as 'systems view of failure' since it supports notions of the nature of man and how one sees reality. More specifically, Flood and Carson (1993) posit that: "The nature of perceived reality is inevitably conditioned by our nature as observing systems" (p. 35). This support is grounded in systems thinking as supported by other researchers including Clemson (1984, p. 21) who suggested that "the activity of observing [done by an analyst] ... has some influence on the observed system [system under study]." Moreover, the systems perspective of failure can be used to generate two additional assertions. First, there can be multiple sources of errors leading to a failure in a system. This assertion challenges the traditional mentality that there can only be one source of error that is identifiable using methods such as *Fault Tree Analysis* focused on determination of a root cause. In complex systems, there can be multiple interconnected factors and cannot easily assigned to system elements and components. Second, an underlying paradigm can serve as a source of error identification. Any one given paradigm only offers a partial view of sources of any error. Taken together, as suggested in Figure 1 below, a complete picture emerges for an example of a bridge system failure.

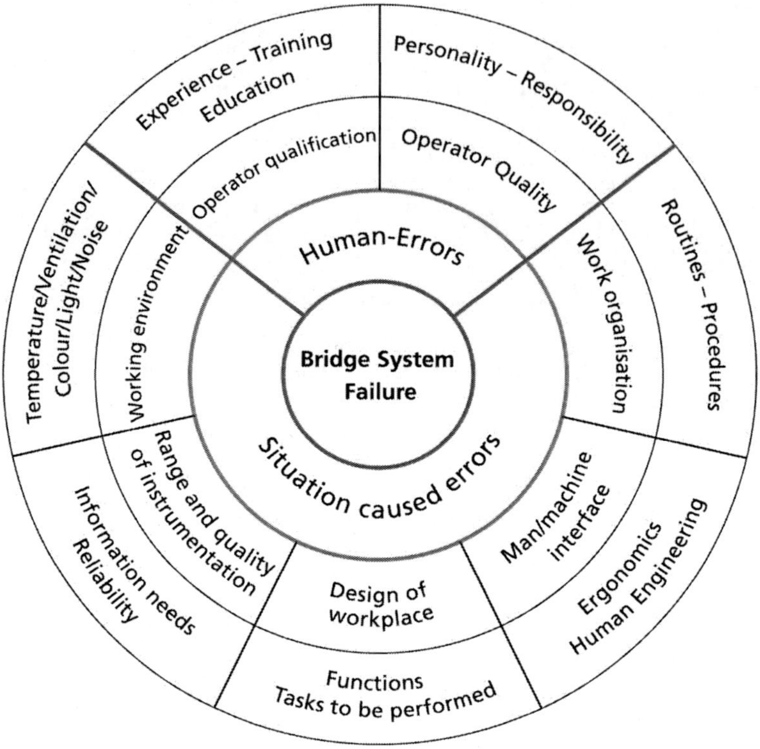


Figure 1. A systems perspective of failure.

This can be illustrated by looking at an example. Consider an example of a Thames Trains service that went through a red signal and collided with a Great Western InterCity Express in the United Kingdom, killing 31 people on October 1999. An interim report suggested that immediate cause of the crash appeared to be a commuter train running a red light. However, it also indicated that the *root causes* were likely to be more complex[1]. In such a situation, it can be useful to understand types of paradigms that influence detection of failures and errors. Table 4 elaborates of failure paradigms.

Notice that each paradigm offers advantages of identifying specific errors. In addition, each paradigm comes with applicability limitations associated with the context within which a system exists. A categorization of complex system failure is suggested in Table 5.

[1] Ladbroke Grove crash report. Retrieved from http://news.bbc.co.uk/2/hi/uk_news/1396613.stm.

**Table 4. Paradigms of Failure**

| Failure Paradigm | Description |
|---|---|
| Formal System Paradigm | Refers to specifications of 'what' the system is and how the system boundaries are defined. Expected failures arise when one's operational notion of the 'system' does not account for what is part of the system and those that are to be considered as part of its environment. |
| Control System Paradigm | Individual system entities are characterized by specific action and reaction relationships. In most cases, these include feedback loops that allows the system to modify its input behavior to achieve a certain system output. Examples of potential engineering failures may occur when these system entity relationships are absent or misrepresented. |
| Communication System Paradigm | Known information channels need to be established for both routine and non-routine operational problems that may arise. When crucial information does not reach intended system entities needed to make operational adjustments, the likelihood of engineering failure is amplified. |
| Engineering and Reliability System Paradigm | Systems rely, to a certain extent, on man-made technology and technical procedures and processes. These are prone to wear and tear and will surely breakdown at some point during the course of system lifetime, whether normally or unexpectedly. |
| Human Factors System Paradigm | Refers to system breakdowns that can be traced back to salient performance attributes of the human/s that are participants of the system. This include discrete components of performance, largely centered around psychological, behavioral, neuromuscular and cognitive time lags inherent with how humans interact with the system. |
| Forecasting and Planning System | Many systems rely on models used to predict short-term or long-term expected outputs of the system components and as a whole. Any deviation away from this assumed parameters in the models is a potential source of future failures. |
| Contextual System Paradigm | A system does not operate in a vacuum. It interacts with its environment in many ways such as transforming inputs from its environments and then producing outputs back for consumption. However, the environment where the system operates may no longer have use for the outputs of the system. The value of the system is fundamentally tied to providing for a specific 'need.' If the need no longer exists, the system falls out of context and will have to be retired or integrated into another system to support a new 'need' altogether. |

**Table 5. Issues in Complex System Failure**

| Complex System Failure | | Elements |
|---|---|---|
| *Hard System Failure* | Failure in system performance based on | Specification/requirements |
| | | Criteria for performance |
| | | Design elements |
| *Soft System Failure* | Failure in system performance due to | Mechanics of system design process |
| | | Dynamics of system design process |
| | | Contextual issues |
| | | Conceptual mismatch |
| | | Expectation mismatch |
| | | Systemic pathologies |

Table 5 is meant to suggest a need for a complementary approach to understanding 'soft' and 'hard' aspects with respect to failure in *systems analysis*. If a complementary approach to systems analysis is undertaken with respect to failures and errors involving complex systems, then, a consideration of several factors, as suggested by the following operand, could be used to inform a more complete systems analysis approach:

$$Complete\ Systems\ Analyis = f\begin{bmatrix}Technical\ requirements, Technical\ perfromance\ measures,\\ Design\ elements\end{bmatrix} + [System\ mechanics, System\ dynamics, Contextual\ issues,\ System\ pathologies, etc.)$$

This operand is meant to evoke the need for consideration of design, deployment, or operational failure. It also involves a consideration of contextual alignment/integration failure, uncompensated patterns that emerge out of interaction of different systems, and the mechanics and dynamic nature of complex system processes. These facets are in addition to traditional elementary consideration of product and service design. Therefore, an analysis based in holistic system failure involves the following aspects:

- Integrated failure with strategic system management on a continuous basis involving pre and post failure analysis.

- The process is systemic in nature (i.e., involving identification, processing, meaning attribution, decision, action, interpretation).
- Consideration of Fail Early, Fail Frequently, Fail Fast, Fail Systemically Deep failure modes to limit the negative consequences experienced from a failure
- It provides a more effective Complex Systems design, deployment, operation, and transformation of the system of interest.

These caveats serve as a reminder that anticipating system failures requires a proactive and holistic system perspective to inform the approach to dealing with any weaknesses or failure points that may occur throughout the system. Certainly, dealing with failures and errors in complex systems involves identification of pathologies. A pathology can be viewed as the condition that acts to reduce system performance. Pathologies may be derived from direct violation of systems theory, including ignoring or inappropriate application of systems theory laws, principles, and theorems (Katina 2015a; Keating and Katina, 2012). The identification of pathologies is a key phase in any methodological approach that invokes an effort commonly referred to as *problem formulation* (Katina, 2015a; Katina, 2015b; Katina, 2016; Keating and Katina, 2016). Problem formulation is considered as a portal into complex system understanding and development of solutions to bring about positive change to a system of interest (Katina, 2016). Pathologies associated with violation of systems theory span an array of circumstances involving dynamics of a system, system goals/missions, information flow, processes, regulation, resources, systemic structures, and understanding (Katina, 2015b).

Recent development in research has suggested the Metasystem Pathologies Method (M-Path Method™ as an approach to identify systems theory-based pathologies. This method consists of five phases: identification, analysis, exploration, systemic implementation, and follow-up as outlined in Table 6 below.

There are several tools that can be deployed to aid in application of the M-Path Method™ efforts to identify pathologies. Specifically, data mining and surveys could be used to ascertain information regarding the presence of pathologies in a system of interest as well as their impact on system operations. Katina (2015b) suggested a Web-based instrument. Fuzzy cognitive mappings can also support development of relationships between pathologies especially when a variety of stakeholders are involved in a system of interest. This is essential since each stakeholder might not share the same values associated with measures related to pathology impact, existence, and implications, even for the same pathology. Using M-Path Method™ approach enables the analyst to incorporate tools and software that can provide substantive information on threads that appears relevant to all stakeholders (Herzing et al. 2014; Katina and Unal, 2015; Mohagheghi, 2015; Papageorgiou and Kontogianni, 2012).

**Table 6. Five Phase of M-Math Method™ for Identification of Pathologies**

| Phase | Description | Products |
|---|---|---|
| *Phase I: Identification* | This phase involves the identification and discovery of the degree to which the systems theory-based pathologies exist in a given situation/system. This phase involves elicitation of information regarding *degree of existence* and *impact* of each pathology. | – Results is an ordinal map that 'bins' pathologies based on levels of existence and potential impact |
| *Phase II: Analysis* | This phase involves an examination of nature and implications of the unique 'landscape' of pathologies for the system interest. This is an attempt to synthetize the data into meaningful information concerning the pathologies. | – An enumerated listing of pathologies<br>– A caption of variability in measures of degree and impact as suggested by participants |
| *Phase III: Exploration* | This phase involves a two-way dialogue between system owners and the analysts involving the *general meaning* of pathologies and *exploration of the meaning in context* for the system of interest. | – A listing of implications for system development<br>– A listing of initiatives to address identified pathologies<br>– A *prioritized enumeration* of pathologies based on feasibility<br>– A set of strategies and corresponding actions designed to impact the identified pathologies |
| *Phase IV: Systemic Implementation* | The purpose of this phase is to ensure that selected responsive strategies are effectively deployed. | – A plan-of-action with specific details on what will be done to actually improve the system along with a time-line |
| *Phase V: Follow-up* | This phase is 'future-focused' and is intended to ensure that the established timeline serves as a place-maker for a re-evaluation of the system by fulfilling two primary purposes: 1) to measure the effects of the strategies/actions as implemented in Phase IV and 2) identification of new pathologies. | – Measures effectiveness of deployment strategies<br>– Articulation of new pathologies<br>– Identifcation of new technologies and strategies against pathologies |

The M-Path Method™, in the context of problem formulation for complex systems, enables one to focus on deep systemic issues that can affect performance of a system. The consideration of pathologies can be part of an integrated approach for analysis of systemic failure identification. It can serve 1) to identify what is wrong in a complex system in terms of pathologies, for an existing system, 2) provide insights into what could go wrong in a system before it comes into existence, or 3) indicate appropriate systemic approaches that could be used to address pathologies.

## CONCLUSION AND FURTHER RESEARCH

The landscape for 21st century systems is characteristically ambiguous, complex, mired with emergent behaviors. This stems from the interdependent nature of systems, which forces systems to operate under conditions of uncertainty. It is from this perspective that one forms the belief that it is essential to have robust methodological approaches to effectively navigate this landscape. The purpose is always to enhance the prospects for continuing system viability (continued existence). Such methodologies as those articulated in this chapter, as well as their foundations of systems theory, can contribute greatly to this pursuit. Emphasis is placed on failure identification as an essential element of complex system analysis. Failure analysis in complex systems forms an integral part of problem formulation and is directly related to identification of systems theory-based pathologies.

A method associated with identification of systems theory-based pathologies (M-Path Method™) provides for this identification as well as corresponding tools. Arguably, the consideration of systems pathologies provides an entry for discovery of deep systemic issues that could affect performance of complex systems. Therefore, the M-Path Method™ offers a critical element for advancing the prospects of systems challenged to operate in the 21st century landscape. More importantly, since systems theory is suggested as a foundation for understanding complex systems, it stands to reason that systems theory provide a basis for identification of complex system failures and errors. In addition, systems theory can be instrumental in the subsequent development of measures and strategies to effectively deal with performance related deficiencies in complex systems.

Finally, authors suggest that the efficacy of any proposed research is typically measured in terms of the results achieved. Thus, future research should include analysis of results based on the deployment of present research, especially the elements pertaining to the identification of pathologies. There is also potential for case applications across a variety of domains including energy, water, transportation, and healthcare. Since systems theory is transdisciplinary, the corresponding methods and its tools derived also exist are domain-free. Case applications might shed more light into pathologies by further validating present research as well as suggesting fruitful paths foreword. The existence of pathologies in different systems might offer generalizable insights into types of pathologies stemming from specific cases. Specific case applications could also serve as the basis for developing tailored strategies and countermeasures for specific problem domains.

## References

Ackoff, R. L. (1974a). *Redesigning the future: A systems approach to societal problems*. New York, NY: John Wiley and Sons.

Ackoff, R. L. (1974b). Systems, messes, and interactive planning. In: *Redesigning the future: Systems Approach to societal problems* (pp. 20-33). New York: NY: John Wiley and Sons Inc.

Ackoff, R. L. (1981a). *Creating the corporate future*. New York, NY: Wiley.

Ackoff, R. L. (1981b). The art and science of mess management. *Interfaces*, 11(1), 20–26.

Adams, K. M., Hester, P. T., Bradley, J. M., Meyers, T. J., and Keating, C. B. (2014). Systems theory as the foundation for understanding systems. *Systems Engineering*, 17(1), 112–123. http://doi.org/10.1002/sys.21255.

Adams, K. M., and Keating, C. B. (2009). *SoSE methodology rev 0.2 (No. NCSoSE Technical Report 009-2009)*. Norfolk, VA: National Centers for System of Systems Engineering.

Adams, K. M., and Keating, C. B. (2011). Overview of the system of systems engineering methodology. *International Journal of System of Systems Engineering*, 2(2/3), 112–119.

Ansoff, H. I. (1984). *Implanting strategic management*. Englewood Cliffs, NJ: Prentice-Hall.

Argyris, C., and Schön, D. (1978). *Organizational learning: A theory of action perspective*. Reading, MA: Addison-Wesley.

Argyris, C., and Schön, D. (1996). *Organizational learning II: Theory, method, and practice*. New York, NY: Addison-Wesley.

Atthill, C. (1975). *Decisions: West oil distribution*. London, UK: P.B. Educational Services.

Beer, S. (1979). *The heart of the enterprise*. New York, NY: John Wiley and Sons.

Beer, S. (1981). *Brain of the firm: The managerial cybernetics of organization*. Chichester, UK: Wiley.

Beer, S. (1985). *Diagnosing the system for organizations*. Oxford, UK: Oxford University Press.

Blanchard, B. S., and Fabrycky, W. J. (2006). *Systems engineering and analysis* ($4^{th}$ ed.). Upper Saddle River, NJ: Pearson - Prentice Hall.

Burrell, G., and Morgan, G. (1979). *Sociological paradigms and organisational analysis*. Burlington, VT: Ashgate Publishing.

Calida, B. Y. (2013). System governance analysis of complex systems (Ph.D.). Old Dominion University, United States -- Virginia.

Checkland, P. B. (1990). Soft systems methodology: A thirty year retrospective. In: P. B. Checkland and J. Scholes (Eds.), *Soft Systems Methodology in Action* (pp. A1-A66). Chichester, UK: John Wiley and Sons Ltd.

Checkland, P. B. (1993). *Systems thinking, systems practice*. New York, NY: John Wiley and Sons.

Checkland, P. B., and Scholes, J. (1990). Soft systems methodology in action. Chichester, UK: John Wiley and Sons Ltd.

Cherns, A. (1976). The principles of sociotechnical design. *Human Relations*, 29(8), 783–792.

Churchman, C. W. (1968). *Challenge to reason*. New York, NY: McGraw-Hill.

Churchman, C. W. (1971). *The design of inquiring systems*. New York, NY: Basic Books.

Churchman, C. W., Ackoff, R. L., and Arnoff, E. (1957). *Introduction to operations research*. New York, NY: Wiley.

Clemson, B. (1984). *Cybernetics: A new management tool*. Tunbridge Wells, Kent, UK: Abacus Press.

Cohen, R. E., and Ahearn, F. L. (1986). *Handbook for mental health care of disaster victims*. In: B. Raphael (Ed.), When disaster strikes. Hutchinson, London: Johns Hopkins University Press.

Componation, P., and Collopy, P. (2012). Systems engineering theory: Addressing complexity and uncertainty in space system architecting. In: *AIAA SPACE 2012 Conference and Exposition*. Pasadena, California: American Institute of Aeronautics and Astronautics. http://doi.org/http://dx.doi.org/10.2514/6.2012-5278.

Conrad, T. P., and Gheorghe, A. V. (2011). Editorial: System of systems engineering in naval application. *International Journal of System of Systems Engineering*, 2(2/3), 89–90.

Digby, J. (1989). *Operations research and systems analysis at RAND, 1948-1967* (No. RAND/N-2936-RC) (p. 4). Santa Monica, CA: RAND.

Fiol, C. M., and Lyles, M. A. (1985). Organizational learning. *The Academy of Management Review*, 10(4), 803–813. http://doi.org/10.2307/258048.

Flood, R. L. (1995). Total systems intervention (TSI): A reconstitution. *Journal of the Operational Research Society*, 46(2), 174–191.

Flood, R. L., and Carson, E. R. (1993). *Dealing with complexity: An introduction to the theory and application of systems science*. New York, NY: Plenum Press.

Flood, R. L., and Jackson, M. C. (1991). *Critical systems thinking: Directed readings*. New York: NY: John Wiley.

Forrester, J. W. (1961). *Industrial dynamics*. Cambridge, MA: MIT press Cambridge, MA.

Gaines, B. R. (1977). Progress in general systems research. In: G. J. Klir (Ed.), *Applied general systems research: Recent development and trends* (pp. 3–28). New York, NY: Plenum Press.

Gibson, J. E., Scherer, W. T., and Gibson, W. F. (2007). *How to do systems analysis*. Hoboken, NJ: Wiley-Interscience.

Hammond, D. (2002). Exploring the genealogy of systems thinking. *Systems Research and Behavioral Science,* 19(5), 429–439. http://doi.org/10.1002/sres.499.

Herzig, S. J. I., Qamar, A., and Paredis, C. J. J. (2014). An approach to identifying inconsistencies in model-based systems engineering. *Procedia Computer Science*, 28, 354–362. http://doi.org/10.1016/j.procs.2014.03.044.

Hester, P. T., and Adams, K. M. (2014). *Systemic thinking: Fundamentals for understanding problems and messes.* New York, NY: Springer Berlin Heidelberg.

INCOSE. (2011). *Systems engineering handbook: A guide for system life cycle processes and activities*. (H. Cecilia, Ed.) (3.2 ed.). San Diego, CA: INCOSE.

Jackson, M. C. (1991). *Systems methodology for the management sciences.* New York, NY: Plenum Press.

Jackson, M. C. (2000). *Systems approaches to management*. New Jersey, NY: Springer.

Jackson, M. C. (2003). *Systems thinking: Creative holism for managers.* Chichester, UK: John Wiley ad Sons Ltd.

Katina, P. F. (2015a). Emerging systems theory–based pathologies for governance of complex systems. *International Journal of System of Systems Engineering*, 6(1/2), 144–159. http://doi.org/10.1504/IJSSE.2015. 068806.

Katina, P. F. (2015b). Systems theory-based construct for identifying metasystem pathologies for complex system governance (Ph.D.). Old Dominion University, United States -- Virginia. Retrieved from http:// search.proquest.com.proxy.lib.odu.edu/docview/1717329758/abstract/29A 520C8C0A744A2PQ/2.

Katina, P. F. (2016). Systems theory as a foundation for discovery of pathologies for complex system problem formulation. In: A. J. Masys (Ed.), *Applications of Systems Thinking and Soft Operations Research in Managing Complexity* (pp. 227–267). Geneva, Switzerland: Springer International Publishing. Retrieved from http://link.springer.com/chapter/ 10.1007/978-3-319-21106-0_11.

Katina, P. F., Keating, C. B., and Jaradat, R. M. (2014a). System requirements engineering in complex situations. *Requirements Engineering*, 19(1), 45–62. http://doi.org/10.1007/s00766-012-0157-0.

Katina, P. F., Pinto, C. A., Bradley, J. M., and Hester, P. T. (2014b). Interdependency-induced risk with applications to healthcare. *International Journal of Critical Infrastructure Protection*, 7(1), 12–26. http://doi.org/10.1016/j.ijcip.2014.01.005.

Katina, P. F., and Unal, R. (2015). Application of fuzzy sets in decision analysis for prioritising critical energy infrastructures. *International Journal of Decision Sciences, Risk and Management,* 6(1), 1–15. http:// doi.org/10.1504/IJDSRM.2015.072762.

Keating, C. B. (2015). Complex system governance: Theory to practice challenges for system of systems engineering. In: *System of Systems Engineering Conference (SoSE)*, 2015 10$^{th}$ (pp. 226–231). http://doi.org/ 10.1109/SYSOSE.2015.7151955.

Keating, C. B., and Bradley, J. M. (2015). Complex system governance reference model. *International Journal of System of Systems Engineering*, 6(1-2), 33–52. http://doi.org/10.1504/IJSSE.2015.068811.

Keating, C. B., and Katina, P. F. (2012). Prevalence of pathologies in systems of systems. *International Journal of System of Systems Engineering*, 3 (3/4), 243–267. http://doi.org/10.1504/IJSSE.2012.052688.

Keating, C. B., and Katina, P. F. (2016). Complex system governance development: A first generation methodology. *International Journal of System of Systems Engineering*, 7(1/2/3), 43–74. http://doi.org/10.1504/IJSSE.2016.076127.

Keating, C. B., Katina, P. F., and Bradley, J. M. (2014). Complex system governance: Concept, challenges, and emerging research. *International Journal of System of Systems Engineering*, 5(3), 263–288.

Keating, C. B., Sousa-Poza, A. A., and Mun, J. (2004). *System of systems engineering methodology*. EMSE: Old Dominion University.

Kovacic, S. F., Sousa-Poza, A., and Keating, C. B. (2007). The National Centers for System of Systems Engineering: A case study on shifting the paradigm for system of systems. *Systems Research Forum*, 2(1), 52–59.

Laszlo, A., and Krippner, S. (1998). Systems theories: Their origins, foundations, and development. In: J. S. Jordan (Ed.), *Systems Theories and A Priori Aspects of Perception* (pp. 47–74). Amsterdam, The Netherlands: Elsevier Science.

Laszlo, E. (1996). *The systems view of the world: A holistic vision for our time*. Cresskill, NJ: Hampton Press.

Martin, J. (2006). *The meaning of the 21st century: A vital blueprint for ensuring our future*. New York, NY: Riverhead Books.

Mason, R. O., and Mitroff, I. I. (1981). *Challenging strategic planning assumptions: Theory, cases, and techniques*. New York, NY: Wiley-Interscience.

Miser, H. J., and Quade, E. S. (1988). *Handbook of systems analysis: Craft issues and procedural choices* (Vol. 2). New York, NY: North-Holland.

Mitroff, I. I., and Emshoff, J. R. (1979). On strategic assumption-making: A dialectical approach to policy and planning. *The Academy of Management Review*, 4(1), 1–12. http://doi.org/10.2307/257398.

Mohagheghi, S. (2015). Fuzzy cognitive maps for identifying fault activation patterns in automation systems. In: E. P. Dadios (Ed.), *Fuzzy Logic - Tool for Getting Accurate Solutions* (pp. 79–99). Rijeka, Croatia: InTech.

Monod, J. (1974). On chance and necessity. In: F. J. Ayala and T. Dobzhansky (Eds.), *Studies in the philosophy of biology* (pp. 357–375). London, UK: Macmillan Press.

Papageorgiou, E., and Kontogianni, A. (2012). Using fuzzy cognitive mapping in environmental decision making and management: A methodological primer and an application. In: S. S. Young and S. E. Silvern (Eds.), *International Perspectives on Global Environmental Change* (pp. 427–450). Rijeka, Croatia: InTech.

Pasmore, W. A. (1988). *Designing effective organizations: The sociotechnical systems perspective*. New York, NY: John Wiley and Sons, Inc.

Rasmussen, J., and Batstone, R. (1989). Why Do Complex Organisational Systems Fail? *World Bank Environmental Working Paper*, No. 20.

Richardson, B. (1994). Socio-technical disasters: Profile and prevalence. *Disaster Prevention and Management: An International Journal*, 3(4), 41–69. http://doi.org/10.1108/09653569410076766.

Rittel, H. W. J., and Webber, M. M. (1973). Dilemmas in a general theory of planning. *Policy Sciences*, 4(2), 155–169. http://doi.org/10.1007/BF01405730.

Skyttner, L. (2005). *General systems theory: Problems, perspectives, practice* (2nd ed.). Singapore: World Scientific Publishing Co. Pte. Ltd.

Smartt, C., and Ferreira, S. (2012). Constructing a general framework for systems engineering strategy. *Systems Engineering*, 15(2), 140–152. http://doi.org/10.1002/sys.20199.

Sousa-Poza, A. A. (2015). Mission engineering. *International Journal of System of Systems Engineering*, 6(3), 161–185. http://doi.org/10.1504/IJSSE.2015.071453.

Sterman, J. D. (2000). *Business dynamics: Systems thinking and modeling for a complex world*. New York, NY: McGraw-Hill.

Strijbos, S. (2010). Systems thinking. In: R. Frodeman, J. T. Klein, and C. Mitcham (Eds.), *The Oxford Handbook of Interdisciplinarity* (pp. 453–470). New York, NY: Oxford University Press, US.

Tainter, J. A. (1988). *The collapse of complex societies*. New York, NY: Cambridge University Press.

Trist, E. L., and Bamforth, K. W. (1951). Some social and psychological consequences of the Longwall Method of coal-getting: An examination of the psychological situation and defences of a work group in relation to the social structure and technological content of the work system. *Human Relations*, 4(1), 3–38. http://doi.org/10.1177/001872675100400101.

Ulrich, W. (1983). *Critical heuristics of social planning: A new approach to practical philosophy*. Bern/Stuttgart: Paul Haupt.

Ulrich, W. (1987). Critical heuristics of social systems design. European *Journal of Operational Research*, 31(3), 276–283. http://doi.org/10.1016/0377-2217(87)90036-1.

von Bertalanffy, L. (1972). The history and status of general systems theory. *Academy of Management Journal*, 15(4), 407–426. http://doi.org/10.2307/255139.

Warfield, J. N. (1976). *Societal systems: Planning, policy and complexity*. New York, NY: Wiley-Interscience.

Weick, K. E. (1988). Enacted sensemaking in crisis situations. *Journal of Management Studies*, 25(4), 305–317. http://doi.org/10.1111/j.1467-6486.1988.tb00039.x.

Whitney, K., Bradley, J. M., Baugh, D. E., and Chesterman, C. W. (2015). Systems theory as a foundation for governance of complex systems. *International Journal of System of Systems Engineering*, 6(1-2), 15–32. http://doi.org/10.1504/IJSSE.2015.068805.

Wilson, B. (1984). *Systems: Concepts, methodologies, and applications*. New York, NY: John Wiley and Sons, Inc.

## BIOGRAPHICAL SKETCH

**Name: Behnido Calida, Ph.D.**

Affiliation: Virginia Tech Transportation Institute

Education: Ph.D. in Engineering Management

Research and Professional Experience:

Professional Appointments:

Senior Project Associate/Data Analyst, Virginia Tech Transportation Institute, Center for Automotive Safety Research, Blacksburg, VA (2014-Present)

Post-doctoral Research Scientist/Adjunct Faculty, National Centers for System of Systems Engineering, Norfolk, VA (2007-2014)

Product Engineer, Operations STMicroelectronics Asia Pacific Pte. Ltd., Singapore, Singapore (2004-2007)

**Honors:**

The Honor Society of Phi Kappa – Member, since 2011

Epsilon Mu Eta – The Engineering Management Honor Society – Member, since 2009.

**Publications Last 3 Years:**

**Book chapters:**

Keating, C.B., Calida, B., Jaradat, R., Katina, P. (2016) Systems Thinking. In: J. Gandhi and J. Farr (Eds.), *The Engineering Management Handbook*, 2nd edition (pp. 245-281). US: ASEM (Forthcoming).

Calida, B., Keating, C. (2013) System Governance: Emergence of Practical Perspectives across the Disciplines. In: A.V. Gheorghe, et al. (Eds.), *Infranomics: Sustainability, Engineering Design and Governance* (pp. 269-296). NY, US: Springer.

Calida, B., Gheorghe, A.V., Unal, R., Vamanu, D. and Radu, V. (2013) Complexity Induced Vulnerability Assessment: How resilient are our academic programs?. In: A.V. Gheorghe, et al. (eds.), *Infranomics: Sustainability, Engineering Design And Governance* (pp. 377-394). NY, US: Springer.

**Journal articles (peer-reviewed)**

Calida, B., Jaradat, R., Abutabenjeh, S., and Keating, C.B., (2016). 'Governance in Systems of Systems: A Systems-based model.' *Int. J. of System of Systems Engineering* (Forthcoming).

Calida, B. (2016). 'Complex system governance: Moving diverse theory to practice,' *Int. J. of System of Systems Engineering*. Vol. 7, Nos. 1/2/3, pp. 22-42.

Calida, B. and Katina, P. (2015). 'Modeling the 2008 financial economic crisis: Triggers, Perspectives and implications from systems dynamics,' *Int. J. of System of Systems Engineering*, Vol. 6, No. 4, pp. 273-301.

Katina, P., Despotou, George, and Calida, B. (2014). 'Sustainability of System of systems,' *Int. J. of System of Systems Engineering*, Vol. 5, No. 2, pp. 93-113.

**Name: Polinpapilinho Katina, Ph.D.**

Affiliation: National Centers for System of Systems Engineering,

Education: Ph.D. (Engineering Management and Systems Engineering)

Address: Department of Engineering Management and Systems Engineering

Old Dominion University

2101-N Engineering Systems Building, Norfolk, VA 23529

**Research and Professional Experience:**

*Systems Thinking: Concepts and Tools for Technical Professionals*, May 2016, College of Continuing Education and Professional Development, Old Dominion University.

*Towards a Method for Identifying Complex System Constraints Impacting Performance of NASA Aeronautics Missions*, 04/2012-04/2013, National Aeronautics and Space Administration/Virginia Space Grant Consortium.

**Professional Appointments:**

Postdoctoral Researcher, September 2015-Present, The National Centers for System of Systems Engineering (NCSOSE), Department of Engineering Management and Systems Engineering, Frank Batten College of Engineering and Technology, Old Dominion University, 2101N Engineering Systems Building, Norfolk, VA 23529

Visiting Scholar, July 2015, Joint Research Center of the European Commission, Institute for Energy and Transport, Westerduinweg 3, NL-1755 LE Petten, the Netherlands

Visiting Research Fellow, June 2015 – September 2015, Energy Department - Nuclear Division, Laboratory of Signal Analysis and Risk Analysis (LASAR), Politecnico di Milano, Via Ponzio 34/3 - 20133, Milano, Italy

**Honors:**

2015 ISERC Best Track Papers Awards: *Systems Engineering*, IIE Industrial and Systems Engineering Research Conference

NSBE Graduate Student Professional Conference Travel Scholarships 2014-2015, National Society of Black Engineers

Bristol Who's Who, 641 Lexington Ave., 15th Floor, New York, NY

Ford Foundation Dissertation Honorable Mention, National Research Council of the National Academies

Nomination for Young Scientists Summer Program (YSSP)/The International Institute for Applied Systems Analysis (IIASA), Laxenburg, Austria

ILR Graduate Fellowship, Institute for Learning in Retirement at Old Dominion University

Love of Learning Award, The Honor Society of Phi Kappa Phi

2012-2013 Ford Foundation Pre-Doctoral Honorable Mention, National Research Council of the National Academies

*Publications Last 3 Years*:

**Books (2)**

Gheorghe, A.V., Masera, M., and Katina, P.F. (Eds.). (2014). *Infranomics: Sustainability, engineering design and governance.* Vol. 24. Geneva, Switzerland: Springer International Publishing [422 pages; guest editors; 60 authors].

Vamanu, B.I., Gheorghe, A.V., and Katina, P.F. (2016). "Critical infrastructures: Risk and vulnerability assessment in transportation of dangerous goods," Cham, Switzerland: Springer International Publishing.

**Book Chapters (3)**

Katina, P.F. (2016). "Systems theory as a foundation for discovery of pathologies for complex system problem formulation," In: Masys, A.J. (Ed.), *Applications of systems thinking and soft operations research in managing complexity* (pp. 227-67). Geneva, Switzerland: Springer International Publishing.

Keating, C.B., Calida, B.Y., Jaradat, R., and Katina, P.F. (*in press*) "Systems thinking," To appear in Merino, D.N. and Farr, J.V. (Eds.), *ASEM Handbook on Engineering Management, 2nd Edition* (pp. xx-xx). Huntsville, AL: ASEM Press.

Katina, P.F. (*in press*). "Individual and Societal Risk (RiskIS): Beyond probability and consequence during Hurricane Katrina," To appear in Masys, A.J. (Ed.), *Disaster Forensics: Understanding root cause and complex causality* (pp. xx-xx). Geneva, Switzerland: Springer International Publishing.

**Journal Publications – Peer Reviewed (16)**

Keating, C.B. and Katina, P.F. (2016). "Complex system governance development: A first generation methodology," In: *International Journal of System of Systems Engineering*, Vol. 7, Nos. 1/2/3, pp. 43-74.

Calida, B.Y. and Katina, P.F. (2015). "Modeling the 2008 financial economic crisis: Triggers, Perspectives and implications from systems dynamics," In: *International Journal of System of Systems Engineering*, Vol. 6, No. 4, pp. 273-301.

Katina, P.F. and Keating, C.B. (2015). "Critical infrastructures: A perspective from systems of systems," In: *International Journal of Critical Infrastructures*, Vol. 11, No. 4, pp. 316-344.

Katina, P.F. and Unal, R. (2015). "Application of fuzzy sets in decision analysis for prioritizing critical energy infrastructures," In: *International Journal Decision Sciences, Risk and Management*, Vol. 6, No. 1, pp. 1-15.

Keating, C.B. and Katina, P.F. (2015). "Editorial: Foundational perspectives for the emerging complex system governance field," In: *International Journal of System of Systems Engineering*, Vol. 6, Nos. ½, pp. 1-14 [guest editors; 150 pages; 14 authors].

Katina, P.F. (2015). "Emerging systems theory-based pathologies for governance of complex systems," In: *International Journal of System of Systems Engineering*, Vol. 6, Nos. ½, pp. 114-159.

Gheorghe, A.V. and Katina, P.F. (2014). "Editorial: Resiliency and engineering systems - Research trends and challenges," In: *International Journal of Critical Infrastructures*, Vol. 10, Nos. ¾, pp. 193-199 [guest editors; 212 pages; 27 authors].

Keating, C.B., Katina, P.F., and Bradley, J.M. (2014). "Complex system governance: concept, challenges, and emerging research," In: *International Journal of System of Systems Engineering*, Vol. 5, No. 3, pp. 263-288.

Katina, P.F., Despotou, G., Calida, B.Y., Kholodkov, T., and Keating, C.B. (2014). "Sustainability of systems of systems," In: *International Journal of System of Systems Engineering*, Vol. 5, No. 2, pp. 93-113.

Katina, P.F., Keating, C.B., and Jaradat, R.M. (2014). "System requirements engineering in complex situations," In: *Requirements Engineering*. Vol. 19, No. 1, pp. 45-62.

Katina, P.F., Pinto, C.A., Bradley, J.M., and Hester, P.T. (2014). "Interdependency-induced risk with application to healthcare," In: *International Journal of Critical Infrastructure Protection*. Vol. 7, No. 1, pp. 12-26.

Reviewed by: Prof. Dr. Charles B. Keating: ckeating@odu.edu.

# INDEX


## #

20th century, 35
21st century, 106, 121, 126

## A

abstraction, x, 69, 71, 78, 99, 106
adaptability, 64
aluminum oxide, 56
Ambiguity, 107
amplitude, 13
anatomy, 77
annealing, 39, 54, 58
Artisan, 74, 76, 87, 90, 95
automated measuring system, vii, 1
automation, 109, 126

## B

bandwidth, 7
barriers, 108
benefits, ix, 40, 105, 112
bias, 18, 20, 74
biological systems, 109
biosensors, 42, 45, 60
birefringence, 8, 34, 37, 48
blood, 60, 78, 81, 85, 87, 95, 99
blood pressure, 85, 87, 95
broadband, 4, 10, 44
browser, 92
business model, 68

## C

$Ca^{2+}$, 40
calibration, 96
categorization, 67, 69, 110, 116
cation, 36, 37
causality, 131
challenges, 64, 100, 102, 115, 125, 126, 132
chemical, 34, 35, 41, 42, 56
chemical vapor deposition, 34
chromatic dispersion (CD), v, vii, viii, 1, 2, 3, 17
Chromatic Dispersion Characterization System, v, 1
circulation, 86, 89
classification, ix, 67, 70, 105, 107, 115
clusters, 41
cognitive function, 66
cognitive map, 119, 126
collaborative development, ix, 64, 73
commercial, viii, 51, 57, 73, 74, 101
communication, ix, 2, 10, 64, 71
compatibility, 34, 67
complement, viii, 33
complex organizations, 112

complexity, 6, 64, 71, 106, 107, 109, 111, 112, 123, 124, 125, 128, 129, 131
composition, 54, 77, 78, 80, 82, 83
computation, 28
computer, 44, 68
computerization, 109
conceptual model, 100
conceptual modelling, 100
conceptualization, ix, 69, 106
configuration, 42, 76
consumption, 117
coordination, 108, 114
cost, viii, 6, 10, 33, 53, 66, 67, 83, 112
covalent bonding, 40
creativity, 71
crystalline, 34, 48, 49

## D

data mining, 119
decomposition, 37, 41, 55, 72, 73, 78, 99
decoration, 34
deficiencies, 121
degradation, 108
deployments, 47
deposition, 34, 41, 56, 59
depth, 40, 41, 108
detection, 2, 20, 45, 58, 59, 116
detection techniques, 58
dielectric constant, 42
diffraction, 52, 58, 59
diffusion, 38, 40, 41, 54, 55
direct measure, 66
discontinuity, 49
dispersion, vii, viii, 1, 2, 6, 10, 17, 24, 28, 29, 30, 40
distribution, 41, 69, 122

## E

economic crisis, 129, 131
electric field, 40, 41, 44
electro optical modulator, vii, viii, 1, 6, 8, 17, 19
electro optical modulator (EOM), vii, viii, 1, 6, 8, 10, 17, 18, 19, 20, 24, 27, 28, 29
electrodes, 11, 18, 27
electromagnetic, ix, 42, 51, 52, 58, 59, 60
electromagnetism, 14
electron, 37, 56, 57
*Emergence*, 107, 129
energy, 52, 58, 59, 60, 89, 122, 125, 132
engineering, vii, ix, 14, 64, 65, 68, 69, 70, 72, 74, 75, 83, 100, 102, 105, 106, 107, 112, 113, 117, 122, 123, 124, 125, 126, 127, 131, 132
environment, ix, 40, 45, 60, 63, 64, 68, 69, 73, 74, 108, 117
environmental degradation, 108
epistemology, 109, 110
etching, 34, 43, 56
European Commission, 130
evaporation, 40
evidence, 73
evolution, 69, 114
execution, 84, 89, 114
exploitation, 42
exposure, 56
extraction, 4, 6

## F

fabrication, viii, 33, 34, 35, 36, 37, 39, 41, 42, 43, 44, 46, 47, 48, 49, 51, 53, 55, 56, 59, 61
failure paradigms, ix, 105, 116
fiber, vii, viii, 1, 2, 3, 4, 5, 6, 7, 10, 14, 15, 17, 20, 24, 27, 28, 29, 34, 35, 44, 45, 47, 60
fiber optics, 14, 15
financial, 129, 131
first generation, 126, 131
fluorescence, 60, 61
forecasting, 47
formation, 42, 47, 55, 57, 58, 59

foundations, 108, 121, 126
fuzzy sets, 125, 132

## G

geometry, 34, 41
glass, vii, viii, 33, 34, 35, 36, 37, 38, 39, 40, 41, 43, 46, 47, 48, 49, 51, 52, 53, 54, 55, 56, 57, 58, 59, 60, 61
governance, 111, 123, 125, 126, 128, 129, 131, 132

## H

hard systems, 106, 110
hard systems thinking, 106
health, 109, 123
health care, 123
heart block, 80
House, 49, 60, 102
human, ix, 60, 64, 65, 67, 76, 77, 80, 95, 110, 111, 114, 117
human body, ix, 64, 76, 77, 95
human values, 110
Hurricane Katrina, 131
hybrid, 45, 48, 49
hydrolysis, 34

## I

idealization, 113
identification, 76, 112, 115, 119, 120, 121, 122
imports, 77
impurities, 39
individuals, 67
industry, 3, 73, 74
insertion, 7, 54
integrated optics, viii, 34, 36, 49, 51, 53, 60
integration, 34, 73, 74, 114, 118
integrity, 65
Interdependence, 107, 108
interface, 7, 36, 37, 39, 41, 42, 69, 78, 81
interference, 3
interoperability, 74, 99
intervention, 111, 114, 124
ion-exchange, vii, viii, 33, 34, 35, 37, 38, 39, 40, 41, 44, 46, 47, 48, 49, 52, 54, 57, 60, 61
Ion-Exchange Technology, v, 33
ions, 34, 36, 37, 40, 41, 44, 53, 54, 55, 57, 58, 59
isolation, 67, 108
issues, 67, 73, 109, 110, 113, 118, 121, 126

## K

$K^+$, 34, 35, 36, 37, 38, 40, 48, 49, 59, 61

## L

landscape, 107, 111, 115, 120, 121
languages, vii, ix, 64, 65, 72, 73, 77, 100
lasers, 14, 45, 46
laws, 109, 110, 119
learning, 111, 113, 122, 124
life cycle, 65, 69, 100, 124
lifetime, 89, 90, 117
light, vii, 1, 2, 6, 7, 10, 11, 19, 34, 42, 44, 52, 56, 58, 116, 122
lithium, 10, 48
low temperatures, 41

## M

magnitude, 111
majority, 93, 99
Malaysia, 1, 17, 33, 51
management, 75, 76, 91, 112, 113, 118, 122, 123, 124, 126
manipulation, 52
manufacturing, 47
mapping, 66, 93, 126
materials, 35, 40, 59, 61

matrix, viii, 34, 40, 51, 53, 93, 96, 97
matter, 14, 89, 100, 115
measurement, vii, viii, 4, 5, 6, 13, 15, 17, 18, 20, 28, 29, 30, 31, 48, 66, 87, 90, 91, 94
melt, 34, 35, 36, 37, 39, 40, 41, 54, 55, 56, 59
melting, 37, 40, 41, 55
metal nanoparticles, ix, 51, 52, 58, 60
methodology, ix, 63, 71, 72, 76, 100, 101, 102, 105, 106, 107, 110, 111, 112, 113, 114, 122, 123, 124, 126, 131
microscope, 44, 54, 58
migration, 44, 46
Model-Based Systems Engineering (MBSE), vii, ix, 63, 64, 65, 69, 70, 71, 72, 73, 74, 100, 101, 102
modelling, vii, ix, 34, 64, 65, 66, 67, 68, 69, 70, 71, 72, 73, 74, 83, 84, 85, 91, 92, 100
models, 65, 66, 67, 68, 69, 70, 71, 74, 75, 84, 93, 99, 100, 102, 117
modern science, 65
Modulation Phase Shift, v, vii, 1, 2, 17
modules, 99
modulus, 65
molecules, 53
momentum, 108
M-Path Method, x, 106, 119, 121

**N**

$Na^+$, 34, 35, 37, 38, 40, 41, 46, 48, 49, 53, 55, 57, 61
Nano particle, 52
nanometer, 57
nanoparticles, ix, 51, 52, 55, 56, 57, 58, 59, 60
nanophotonics, 58
nanostructures, 52, 58
nanowires, 52
National Aeronautics and Space Administration, 130
National Research Council, 130
negative consequences, 119
Netherlands, 126, 130
nodes, 86, 89, 99
nucleus, 68

**O**

operations, 78, 112, 119, 123, 131
operations research, 112, 123, 131
optical communications, 34
optical fiber, 2, 3, 10, 15, 34, 35, 42, 47
optical networks, 2
optical properties, viii, 10, 51, 53, 60
optical sensors, 33, 42, 52
optical waveguide, vii, viii, 10, 33, 34, 35, 36, 44, 46, 47, 48, 52, 54, 56
Optical Waveguide Sensor, v, vii, 33, 35
optimization, vii, viii, 17, 111, 114

**P**

participants, 117, 120
partition, 89, 92
pathologies, x, 106, 118, 119, 120, 121, 122, 125, 131
pathology, 119, 120
photolithography, 43, 54, 56
photonics, 49, 52, 59, 61
physical properties, 2, 99
physics, 48, 60, 61, 109
platform, viii, 51, 100
polarizability, 34, 37, 38
polarization, 2, 6, 8, 9, 34, 40
policy, 113, 126, 128
potassium, 34, 39, 54, 57
principles, viii, 1, 65, 69, 72, 76, 109, 110, 119, 123
probability, 83, 115, 131
probability distribution, 83
*problem formulation*, 106, 111, 119, 121, 125, 131
problem solving, ix, 105
programmability, 12

programmable digital oscilloscope (DOSC), viii, 1, 2
programming languages, 72
propagation, 33, 34, 37, 39, 52, 61
propagation loss, 33, 37, 39, 52
prototypes, 69, 70
psychology, 109

## Q

quality of service, 3

## R

radius, 8, 34, 37, 38
Raman spectroscopy, 52, 58, 59, 61
reactions, 42
reality, 66, 69, 70, 71, 109, 115
reasoning, 66
reception, 78
recognition, 114
reductionism, 108, 109
refractive index, 2, 27, 33, 34, 36, 37, 40, 41, 52, 53, 54, 55
reliability, 64, 67, 75, 100
requirements, 4, 40, 41, 42, 58, 66, 68, 69, 73, 74, 75, 84, 91, 92, 93, 95, 96, 99, 111, 115, 118, 125, 132
researchers, 36, 40, 108, 111, 115
resolution, 8, 12
resonator, 45, 46
resources, 106, 119
response, 2, 10, 19, 24, 28, 52, 75, 90, 108, 113
retardation, 8, 9
RF frequencies, vii, viii, 17
RF signal, viii, 1, 2, 6, 13, 14, 18, 19, 20, 21, 22, 23, 24, 25, 26, 27, 28
risk, 65, 92, 112, 125, 132
root, 72, 115, 116, 131

## S

salts, 37, 41
scattering, 40, 52, 55, 56, 60, 61
science, 61, 65, 71, 106, 108, 109, 122, 124
scientific method, 108
self-regulation, 112
semantics, 71, 84
semiconductors, 34
sensing, 34, 42, 44, 46, 52, 58, 61
sensitivity, 35, 83
sensors, vii, 33, 35, 39, 42, 43, 49, 45, 52, 53, 58, 60
sequencing, 87
services, 40, 64, 78, 89
set theory, 70
shape, 19, 24, 27, 34, 41, 60, 67
shortage, 106, 110, 111
showing, 4, 27, 30, 73, 77, 80, 91
signals, vii, 1, 4, 5, 7, 11, 13, 24, 27, 78
silver, vii, 34, 35, 37, 39, 40, 41, 44, 47, 49, 56, 59, 60, 61
simulation, 67, 70, 102
sine wave, 12
Singapore, 127, 128
single-loop learning, 113
SiO2, 11, 47
smoothness, 54
social sciences, 109
social structure, 127
society, 64, 67, 109, 113
sodium, 40, 41, 54, 56, 60
soft systems, 106, 110
soft systems thinking, 106, 110
software, vii, 1, 6, 7, 64, 65, 67, 68, 72, 73, 74, 75, 78, 99, 102, 109, 120
sol-gel, 34
solution, 44, 109, 111
specialists, 108
specialization, 77, 83
specifications, 69, 93, 117
spectral component, 3
spectroscopy, 14, 52, 58, 59, 60, 61
stability, 12, 35, 38
stakeholders, ix, 63, 64, 66, 70, 71, 73, 74, 77, 91, 100, 110, 119
standardization, 65

state, ix, 8, 9, 35, 40, 48, 64, 66, 68, 70, 71, 73, 75, 84, 90, 91, 94, 107, 115
strategic management, 122
strategic planning, 126
stress, 34, 35
structure, 10, 36, 39, 40, 67, 68, 70, 71, 72, 74, 75, 76, 77, 78, 81, 89, 92, 99, 106, 107, 112, 113, 127
subjectivity, 110, 111, 115
substrate, viii, 11, 33, 34, 36, 39, 40, 41, 42, 46, 53, 54, 55, 56
sustainability, 64, 100
Switzerland, 125, 131
SysML, v, vii, ix, 63, 64, 65, 72, 73, 74, 75, 76, 77, 79, 83, 84, 90, 91, 92, 99, 100, 101, 102, 103
system analysis, vii, 121
system model, 64, 73, 75, 91
systemic failure, vii, ix, 106, 107, 111, 115, 121
systems engineering, i, ii, iii, v, vii, ix, 63, 64, 65, 70, 72, 73, 74, 83, 100, 101, 102, 103, 105, 106, 112, 113, 122, 124, 125, 126, 127, 128, 129, 130, 131, 132
Systems Philosophy, 109, 110
*Systems Science*, 103, 109, 110
Systems Technology, 109, 110
systems theory, x, 106, 109, 110, 111, 119, 120, 121, 122, 125, 127, 132
systems theory-based methodologies, 106, 111
systems theory-based pathologies, 106, 119, 120, 121, 132
systems-based methodologies, 110, 111, 112, 114

## T

target, 66, 67, 70, 100
teams, 76, 100, 101
technician, 77
techniques, 15, 42, 44, 58, 68, 71, 84, 126
technology(ies), vii, viii, 33, 34, 35, 46, 47, 51, 61, 109, 117, 120
temperature, 7, 8, 36, 37, 40, 41, 44, 54, 55
testing, 15, 67
thermal stability, 38
three-dimensional representation, 70
time lags, 117
titanium, 10, 11
top-down, 72
trade, 5, 72, 73, 83, 99
transformation, 84, 119
transmission, viii, 2, 6, 10, 11, 18, 19, 20, 33, 45, 58
transparency, 10, 53
transportation, 122, 131
tunable laser (TLS), vii, 1, 2

## U

Uncertainty, 107
UV light, 56

## V

validation, 67, 69, 73
variables, 111, 112
variety of domains, 122
vision, 70, 71, 126
visualization, 70
vulnerability, 131

## W

waveguide technology, vii, viii, 47, 51, 61
wavelengths, vii, viii, 3, 4, 5, 9, 17, 30, 39